I0034410

Nature-Inspired Optimization in Advanced Manufacturing Processes and Systems

Artificial Intelligence (AI) in Engineering

Series Editors:
Kaushik Kumar and J. Paulo Davim

This new series will target artificial intelligence (AI) applied in engineering disciplines, providing a collection of textbooks and research books. The series of books will provide an understanding of AI using a simple language and incorporate tools and applications to assist in the learning process. The books that will be published in this series will focus on areas such as artificial intelligence and philosophy, applications of AI in mechatronics, AI in automation, AI in manufacturing, AI and Industry 4.0, cognitive aspects of AI, intelligent robotics, smart robots, cobots, machine learning, conscious computers and intelligent machines.

Nature-Inspired Optimization in Advanced Manufacturing
Processes and Systems
Edited by Ganesh M. Kakandikar and Dinesh G. Thakur

For more information on this series, please visit: www.routledge.com/Artificial-Intelligence-AI-in-Engineering/book-series/CRCAIIE

Nature-Inspired Optimization in Advanced Manufacturing Processes and Systems

Edited by
Ganesh M. Kakandikar and Dinesh G. Thakur

CRC Press
Taylor & Francis Group
Boca Raton London New York

CRC Press is an imprint of the
Taylor & Francis Group, an **informa** business

MATLAB® is a trademark of The MathWorks, Inc. and is used with permission. The MathWorks does not warrant the accuracy of the text or exercises in this book. This book's use or discussion of MATLAB® software or related products does not constitute endorsement or sponsorship by The MathWorks of a particular pedagogical approach or particular use of the MATLAB® software

First edition published 2021
by CRC Press
6000 Broken Sound Parkway NW, Suite 300, Boca Raton, FL 33487-2742

and by CRC Press
2 Park Square, Milton Park, Abingdon, Oxon, OX14 4RN

© 2021 Taylor & Francis Group, LLC

CRC Press is an imprint of Taylor & Francis Group, LLC

Reasonable efforts have been made to publish reliable data and information, but the author and publisher cannot assume responsibility for the validity of all materials or the consequences of their use. The authors and publishers have attempted to trace the copyright holders of all material reproduced in this publication and apologize to copyright holders if permission to publish in this form has not been obtained. If any copyright material has not been acknowledged please write and let us know so we may rectify in any future reprint.

Except as permitted under U.S. Copyright Law, no part of this book may be reprinted, reproduced, transmitted, or utilized in any form by any electronic, mechanical, or other means, now known or hereafter invented, including photocopying, microfilming, and recording, or in any information storage or retrieval system, without written permission from the publishers.

For permission to photocopy or use material electronically from this work, access www.copyright. com or contact the Copyright Clearance Center, Inc. (CCC), 222 Rosewood Drive, Danvers, MA 01923, 978-750-8400. For works that are not available on CCC please contact mpkbookspermissions@ tandf.co.uk

Trademark notice: Product or corporate names may be trademarks or registered trademarks, and are used only for identification and explanation without intent to infringe.

Library of Congress Cataloging-in-Publication Data
Names: Kakandikar, Ganesh M., editor. | Thakur, Dinesh G., editor.
Title: Nature-inspired optimization in advanced manufacturing processes and systems / edited by Ganesh M. Kakandikar and Dinesh G. Thakur.
Description: First edition. | Boca Raton : CRC Press, 2020. | Series: Artificial intelligence (AI) in engineering | Includes bibliographical references and index.
Identifiers: LCCN 2020028003 (print) | LCCN 2020028004 (ebook) | ISBN 9780367532604 (hardback) | ISBN 9781003081166 (ebook)
Subjects: LCSH: Industrial engineering—Data processing. | Natural computation—Industrial applications.
Classification: LCC T57.5 .N38 2020 (print) | LCC T57.5 (ebook) | DDC 670.285/6382—dc23
LC record available at https://lccn.loc.gov/2020028003
LC ebook record available at https://lccn.loc.gov/2020028004

ISBN: 978-0-367-53260-4 (hbk)
ISBN: 978-1-003-08116-6 (ebk)

Typeset in Times
by codeMantra

Contents

Foreword

It is a great honour to write the foreword for the first edition of the book entitled *Nature Inspired Optimization in Advanced Manufacturing Processes and Systems* under the series *Artificial Intelligence (AI) In Engineering* from CRC Press. In the present era of global economic crisis, manufacturing sectors have to produce products not only with the least cost but also with the aim of enhanced quality and productivity. The adoption of advanced manufacturing technologies and the application of optimization towards the same is the call for the day as the only viable solution.

I congratulate the editors who, I feel, were successful in accomplishing the Herculean task of gathering, filtering and assembling the relevant and important chapters contributed by eminent researchers in the field of optimization in the domain of manufacturing with a wide range of topics from conventional to non-conventional manufacturing techniques applying statistical, heuristic and metaheuristic optimization methods.

It is my hope and expectation that this book will provide an effective learning experience and referenced resource for all researchers, students and professionals from the manufacturing fraternity to support the mankind to come out of the economical crunch due to the current COVID-19 pandemic.

Kaushik Kumar
Series Editor
Artificial Intelligence (AI) In Engineering

Preface

In the present era, organizations in general and manufacturing industries in particular are increasingly becoming more cost-conscious and aiming at enhanced quality of products and productivity of the system. Organizations are finding themselves to be strategically disadvantaged, if they are confined to only products without considering the optimized way of producing it. The adoption of advanced manufacturing technologies and the changing boundaries of the firm have brought the importance of optimization. This is because optimization is the only tool that helps organizations realize their objectives of production to be met and helps managers take adequate decisions. In the business model of the past, production in general to a large extent was confined to a process without considering the optimized way of utilizing the same for enhancing the productivity at a lower cost. Significant research activities have taken place in the area of optimization in the recent years. However, there exists a lack of representative case studies for understanding the optimization techniques in general and nature-inspired optimization techniques in particular in advanced manufacturing processes and systems. The Editors are pleased to present the book entitled *Nature Inspired Optimization in Advanced Manufacturing Processes and Systems.*

The contributing authors are researchers in the domain of manufacturing optimization. The book covers a wide range of topics, including the optimization of electric discharge machining, friction stir welding, incremental sheet forming, electrochemical-based machining processes, drilling process, micro-hole drilling, burnishing process and abrasive water jet machining. It also covers the research work on systems such as efficient cluster head selection for manufacturing processes, power take-off gear box, availability optimization of thermal power plants and manufacturing tolerance for selective assembly techniques. Various nature-inspired optimization techniques have been utilized, such as genetic algorithms, Jaya algorithm, artificial neural networks, krill herd algorithm, civilized swarm optimization, intelligent water drops algorithm and artificial bee colony algorithm. We believe that the book will serve as a reference for all the researchers in the domain of manufacturing processes and systems.

Dr. Ganesh M. Kakandikar
Dr. Dinesh G. Thakur

MATLAB® is a registered trademark of The MathWorks, Inc. For product information, please contact:

The MathWorks, Inc.
3 Apple Hill Drive
Natick, MA 01760-2098 USA
Tel: 508-647-7000
Fax: 508-647-7001
E-mail: info@mathworks.com
Web: www.mathworks.com

Editors

Ganesh M. Kakandikar is a Head Associate Professor at the School of Mechanical Engineering at Dr. Vishwanath Karad MIT World Peace University, Pune. He completed a Ph.D. in Mechanical Engineering from Swami Ramanand Teerth Marathwada University, Nanded, in 2014. He has 20 years of experience in teaching, research and administration. He has authored three books published internationally by Lambert Academic and Taylor and Francis Group and contributed book chapters published by Wiley, Springer and Elsevier. Around 65+ publications are at his credit in national/international journals and conferences. He is a reviewer of many reputed journals. He has chaired sessions in various national and international conferences and worked on their advisory committees. His publications have widely been cited by researchers. He is a professional member of various societies, including International Society on Multiple Criteria Decision Making, International Society for Structural and Multidisciplinary Optimization, and Sheet Metal Forming Research Association. He has executed many research projects with industries, especially on optimization of automotive components. He has guided several M. Tech. theses and adjudicated many of other universities. He is a recipient of many academic and industry awards, including Indo Global Engineering Excellence Award 2015 from Indo Global Chamber of Commerce, Industries and Agriculture for Research Work in Engineering, Cooper Engineering Prize for Best Paper in Mechanical Engineering for the paper entitled 'Thinning Optimization in Punch Plate using Flower Pollination Algorithm' at the Annual Technical Paper Meet 2015 of The Institution of Engineers (India) at Pune, and Appreciation Award in ANSYS Hall of Fame 2017. He is an awardee of Young Scientist International Travel Grant by Government of India. He regularly writes columns in Indian print media on education systems and reforms.

Dinesh G. Thakur is presently working as a Professor in Mechanical Engineering at Defence Institute of Advanced Technology (Deemed to be University, DRDO Ministry of Defence), Girinagar, Pune, since 2011. He has more than 20 years of teaching, administrative and research experience. He completed his master's and doctorate from Indian Institute of Technology, Madras. Prior to 2011, he worked with Dr. Babasaheb Ambedkar Technological University, Lonere, Maharashtra. His areas of interest are manufacturing considerations in design, high-speed machining/green machining, metal forming, etc. He has organized 15 short-term training programs. He has delivered 65 invited talks in various institutes. He is a recipient of many awards such as Teacher of the Year Award 2013 at DIAT, Pune, Member of Board of Management, DIAT, Pune, and Member, Academic Council, DIAT, Pune. He has received a best paper award in World Congress on Engineering 2009 at London. He has been listed in Who's Who in the World 2010. He is a member on various academic and management authorities of many universities/institutes. He is a lifetime member of Indian Society for Technical Education, Institution of Engineers,

American Society of Mechanical Engineers, Indian Institute of Metals, Aeronautical Society of India, MRSI, etc. He is a reviewer of many reputed journals. He has guided 35 M. Tech. students and is currently guiding seven students. Four research scholars have been awarded doctorate under his guidance, and four are presently working. He has visited China, Hong Kong, Macau, Singapore, Malaysia, Japan, the UK, Switzerland, etc. He has over 50 publications in reputed journals.

Contributors

Syed Anjum Alam
Department of Mechanical Engineering,
University Teaching Department
Rajasthan Technical University
Kota, India

E. Balasubramanian
Department of Mechanical Engineering,
Centre for Autonomous System
Research
Vel Tech Rangarajan Dr. Sagunthala
R&D Institute of Science and
Technology
Chennai, India

Barani S.
Department of Communication and
Electronics Engineering
Satyabhama Institute of Science
and Technology (Deemed to be
University)
Chennai, India

Anand K. Bewoor
Cummins College of Engineering for
Women
Pune, India

Hrudaya Jyoti Biswal
School of Mechanical Sciences
Indian Institute of Technology
Bhubaneswar
Bhubaneswar, India

K. Chandrasekar
Department of Mechanical Engineering
PSN College of Engineering and
Technology
Tirunelveli, India

G. Chandrasekaran
Department of Mechanical Engineering
Sree Sakthi Engineering College
Coimbatore, India

Sandip Chavan
School of Mechanical Engineering
MIT World Peace University
Pune, India

Manish Dadhich
Department of Mechanical Engineering,
University Teaching Department
Rajasthan Technical University
Kota, India

Mahesh B. Davangeri
Department of Mechanical Engineering
Sahyadri College of Engineering &
Management
Mangalore, India

Vikas Dhawan
Department of Mechanical Engineering,
Faculty of Engineering &
Technology
Shree Guru Gobind Singh Tricentenary
University
Gurugram, India

Pramod D. Ganjewar
School of Computer Engineering and
Technology
MIT Academy of Engineering
Pune, India

Ashish Goyal
Department of Mechanical Engineering
Manipal University Jaipur
Jaipur, India

Ankur Gupta
Department of Mechanical Engineering
Indian Institute of Technology Jodhpur
Jodhpur, India

Hanumant P. Jagtap
Department of Mechanical Engineering
Zeal College of Engineering and
   Research
Pune, India

Ajith G. Joshi
Department of Mechanical Engineering
Canara Engineering College
Bantwal, India

K. Kiran
Department of Mechanical Engineering
Dr. N.G.P Institute of Technology
Coimbatore, India

Ajay Kumar
Department of Mechanical Engineering,
   Faculty of Engineering &
   Technology
Shree Guru Gobind Singh Tricentenary
   University
Gurugram, India

Deepak Kumar
Department of Mechanical Engineering,
   Faculty of Engineering &
   Technology
Shree Guru Gobind Singh Tricentenary
   University
Gurugram, India

Parveen Kumar
Department of Mechanical Engineering
Rawal Institute of Engineering and
   Technology
Faridabad, India

Ravinder Kumar
Lovely Professional University
Phagwara, India

Vijay Kurkute
Department Mechanical Engineering,
   Bharati Vidyapeeth (Deemed to be
   University) College of Engineering
   Pune
Pune, India

N. Lenin
Department of Mechanical
   Engineering
Vel Tech Rangarajan Dr. Sagunthala
   R&D Institute of Science and
   Technology
Chennai, India

M. Manjaiah
Department of Mechanical
   Engineering
National Institute of Technology
Warangal, India

Diwesh B. Meshram
Department of Mechanical Engineering
Central Institute of Plastics Engineering
   and Technology
Korba, India

Mukuna Patrick Mubiayi
Mechanical and Industrial Engineering
    Department
College of Science, Engineering and
    Technology, University of South
    Africa
Johannesburg, South Africa

K. N. Nandurkar
Department of Production
    Engineering
K.K. Wagh Institute of Engineering
    Education and Research
Nashik, India

V. Pandu Ranga
School of Mechanical Sciences
Indian Institute of Technology
    Bhubaneswar
Bhubaneswar, India

Firozkhan Pathan
D. Y. Patil Institute of Technology
Pune, India

P.J. Pawar
Department of Production Engineering
K.K. Wagh Institute of Engineering
    Education and Research
Nashik, India

Yogesh M. Puri
Department of Mechanical Engineering
Visvesvaraya National Institute of
    Technology
Nagpur, India

D. Rajamani
Department of Mechanical Engineering,
    Centre for Autonomous System
    Research
Vel Tech Rangarajan Dr. Sagunthala
    R&D Institute of Science and
    Technology
Chennai, India

Veeredhi Vasudeva Rao
Mechanical and Industrial Engineering
    Department
College of Science, Engineering and
    Technology, University of South
    Africa
Johannesburg, South Africa

K. Ravi Kumar
Department of Mechanical Engineering
KPR Institute of Engineering and
    Technology
Coimbatore, India

Neelesh Kumar Sahu
Department of Mechanical Engineering
Medi-Caps University
Indore, India

R. Saravanan
Department of Mechanical Engineering
Vaigai College of Engineering
Madurai, India

R. S. Shukla
Department of Mechanical Engineering
Sardar Vallabhbhai National Institute of
    Technology
Surat, India

D. Singh
Department of Mechanical Engineering
Sardar Vallabhbhai National Institute of
    Technology
Surat, India

M. Siva Kumar
Department of Mechanical Engineering,
    Centre for Autonomous System
    Research
Vel Tech Rangarajan Dr. Sagunthala
    R&D Institute of Science and
    Technology
Chennai, India

V. S. Sree Balaji
Department of Mechanical Engineering
Roever Engineering College
Perambalur, India

R. Suresh
Department of Mechanical and
    Manufacturing Engineering
M.S. Ramaiah University of Applied
    Sciences
Bengaluru, India

Sanjeev J. Wagh
Department of Information Technology
Government College of Engineering
Karad, India

1 Investigation on Process Parameters of EN-08 Steel by Using DoE and Multi-Objective Genetic Algorithm Approach

Syed Anjum Alam
Rajasthan Technical University

Ashish Goyal
Manipal University Jaipur

Manish Dadhich
Rajasthan Technical University

CONTENTS

1.1   INTRODUCTION

The milling machining process is the most widely used process to fabricate complex profiles on difficult-to-cut materials. Oktem et al. [1] improved the slicing condition of the milling machining by using the response surface methodology approach. Prete et al. [2] built up a forecast model for uneven surface in flat end mill utilizing RSM approach. The ANN methodology was utilized to anticipate surface roughness, and GA was utilized to improve the surface roughness. Alam et al. [3] selected speed, feed rate, and depth of cut parameters of milling machine. The quadratic expectation was combined with GA to enhance the machining procedure parameters for surface roughness. Chandrasekaran et al. [4] surveyed the use of soft computing tools, i.e., neural networks, fuzzy sets, genetic algorithms, simulated annealing, ant colony optimization, and PSO, in various machining processes such as turning, milling, drilling, and grinding. Turkes et al. [5] proposed a model of surface unpleasantness to explore the impacts of geometrical parameters of the cutting device. It was concluded that the quadratic model is suitable for the ideal estimation of geometrical parameters.

Kadirgama et al. [6] built up a model to streamline the machining states of aluminum alloys with carbide-coated inserts using design of experiments and response surface methodology. Patel et al. [7] investigated the impact of different machining parameters during the machining of Al 6351–T6 alloy with four components and five dimensions. The design of experiments methodology was utilized to recognize the critical parameters. Pandey et al. [8] performed experiments on Inconel 600 alloy and developed micro-features for the aerospace applications. The design of experiments (DoE) methodology was used to identify the significant and non-significant parameters. Ahmet et al. [9] studied the impact of process parameters on surface roughness and factor levels with minimum surface roughness in pocket machining. The authors found that the surface roughness relates adversely to cutting speed and positively to feed rate and cutting depth. Goyal et al. [10] used the design of experiments methodology to determine the optimum machining parameters. The experiments were performed using wire EDM. It was observed the DoE approach is most suitable to identify the optimum values of the parameters. The review work shows that there is need to select process parameters of milling processes during the fabrication of complex profiles. The objective of this study is to investigate the effect of process parameters, i.e., spindle speed, feed rate, and depth of cut, on the machining performance, i.e., surface roughness and cutting time. The significant parameters are identified by the design of experiments methodology. The predictive models were created depending on response surface methodology and multi-objective genetic algorithm (MOGA) approach to examine the results and combined effects of process parameters.

1.2   MATERIALS AND METHODOLOGY

The experiments were performed on a vertical machining center. A high-speed steel end mill cutter under dry condition was used for the machining process. The machine is installed at CIPET Jaipur tool room. The milling machine is shown in Figure 1.1.

FIGURE 1.1   Milling machine.

The spindle speed, feed rate, and depth of cut are selected for the experimental work. The surface roughness and cutting time are selected as response parameters. The EN-08 alloy steel is selected as the workpiece material. The EN-08 alloy is widely used for the fabrication of sheet metal items, car parts, and home apparatuses.

A standard ANOVA methodology is used to identify the significant parameters (mean and variation). Interaction graphs are used to select the best combination of interactive parameters. Roughness is a measure of the texture of a surface. Surface roughness (μm) measurements were taken three times by using a Mitutoyo Surftest, a portable surface roughness tester. The average value was considered as surface roughness value. Cutting time is the time taken to cut a material, and it is required to improve productivity. Cutting time is measured by using a stopwatch. Table 1.1 presents the selected process parameters and their level for the present work. The pilot experiments were performed to identify the important process parameters and their level. The purpose of the pilot experiments is to study the effects of the milling process parameters on performance measures. The effects of these input parameters on both response parameters are studied by using one-factor-at-a-time approach. The fabricated specimens are presented in Figure 1.2.

TABLE 1.1
Process Parameters and Their Levels

	Parameter	−1	0	1	Unit
A	Feed rate	250	350	450	mm/min
B	Cutting speed	2,000	3,000	4,000	rpm
C	Depth of cut	0.1	0.2	0.3	mm

FIGURE 1.2　Specimens fabricated by milling process.

1.3　RESULTS AND DISCUSSION

In the present study, two response variables, i.e., cutting time and surface roughness, are selected. A low value of cutting time and surface roughness reduces the cost of the production and also plays an important role in the manufacturing industries. A high value of cutting time and surface roughness increases the cost of production, which is not desirable. The Table 1.2 presents the experimental results for CT and SR.

TABLE 1.2
Experimental Results for CT and SR

Run	Feed Rate (mm/min)	Speed (rpm)	DoC (mm)	Cutting Time (s)	Surface Roughness (Ra)(µm)
1	250	4,000	0.3	2.41	4.65
2	350	3,000	0.2	3.45	4.96
3	350	3,000	0.2	5.38	5.28
4	350	3,000	0.2	5.37	4.88
5	450	4,000	0.3	2.54	4.91
6	350	3,000	0.2	5.35	5.11
7	350	3,000	0.2	5.36	4.75
8	250	2,000	0.3	5.13	5.49
9	250	2,000	0.1	14.25	4.61
10	350	2,000	0.2	5.36	4.69
11	350	3,000	0.2	5.36	5.10
12	450	3,000	0.2	4.22	4.83
13	450	2,000	0.1	4.00	4.84
14	350	4,000	0.2	5.37	4.42
15	450	2,000	0.3	3.00	5.54
16	250	3,000	0.2	7.50	4.99
17	350	3,000	0.1	7.80	5.30
18	350	3,000	0.3	3.45	5.60
19	250	4,000	0.1	15.10	4.84
20	450	4,000	0.1	8.45	4.62

1.3.1 RANK IDENTIFICATION FOR CUTTING TIME (CT)

The rank of the process parameters for cutting time is presented in Table 1.3. As seen in Table 1.3, the delta is calculated for each factor and then the rank is given to each factor. The mean data analysis for cutting time is presented in Figure 1.3. It is seen from the figure that the depth of cut plays the most important role, whereas the spindle speed plays the least important role. It may be due to the very high speed and the mechanical strength of the test piece material.

1.3.2 OPTIMAL SOLUTION FOR CT

The optimal solution is found by using mean data analysis. The optimal solution for cutting time is presented in Table 1.4 with predicted values. The Roy formulation techniques are used to identify the optimal solution for CT.

TABLE 1.3
Rank Identification for CT

Level	Feed Rate	Speed	DoC
1	8.878	6.348	9.92
2	5.405	5.603	5.499
3	4.442	6.774	3.306
Delta	4.436	1.171	6.614
Rank	2	3	1

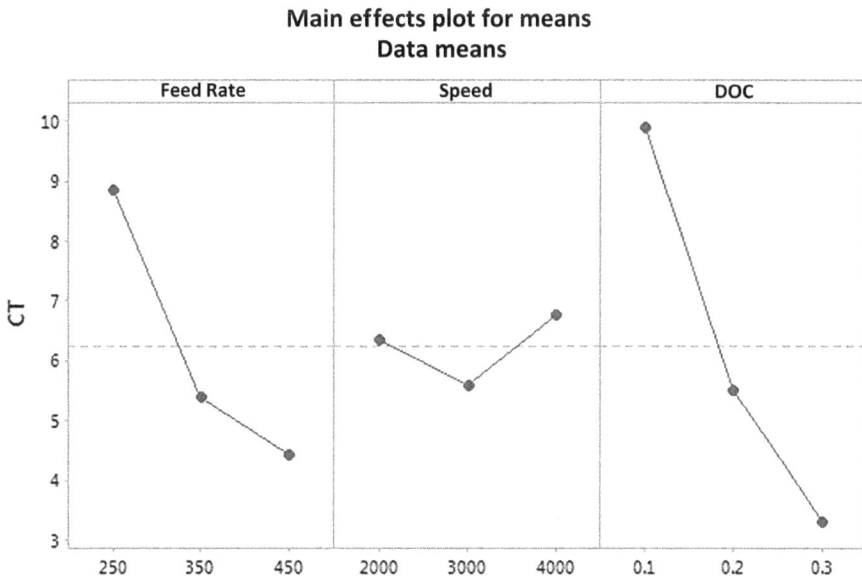

Main effects plot for means
Data means

FIGURE 1.3 Mean data analysis for CT response.

6

Nature-Inspired Optimization in Manufacturing Process

TABLE 1.4
Optimal Solution for Cutting Time (CT)

Response	Feed	Speed	DoC	Org	Prd
CT	450	3,000	0.3	1.75	1.46

The following formula is used for the prediction of optimal solution.

$$\mu_{response} = \bar{A}_{Ln} + \bar{B}_{Ln} + \bar{C}_{Ln} - (F-1)\bar{R} \qquad (1.1)$$

Here, Ln represents level number and F represents the number of factors.

$$\mu_{CT} = \bar{A}_{450} + \bar{B}_{3000} + \bar{C}_{0.3} + -2\bar{R} \qquad (1.2)$$

$$\mu_{CT} = 4.44 + 5.60 + 3.30 - 2*5.94 = 1.46 \qquad (1.3)$$

Now, the experiment for optimal solution is performed on a milling machine to verify the cutting time with the predicted value of cutting time.

1.3.3 RANK IDENTIFICATION FOR SURFACE ROUGHNESS (RA)

Table 1.5 presents the rank identification for SR. The delta is calculated for each factor, and the rank of the parameter is decided. In this case, for minimum surface roughness, spindle speed plays the most important role, whereas the feed rate plays the least important role. Figure 1.4 shows the mean data analysis for surface roughness.

1.3.4 OPTIMAL SOLUTION FOR RA

The optimal solution is found by using mean data analysis for each process parameter, and the results are presented in Table 1.6 with predicted values. The Roy formulation techniques are used to identify the optimal solution for SR.

TABLE 1.5
Rank Identification for Surface Roughness

Level	Feed Rate	Speed	DoC
1	4.916	5.034	4.842
2	5.005	5.147	4.789
3	4.948	4.688	5.238
Delta	0.089	0.459	0.449
Rank	3	1	2

Main effects plot for means
Data means

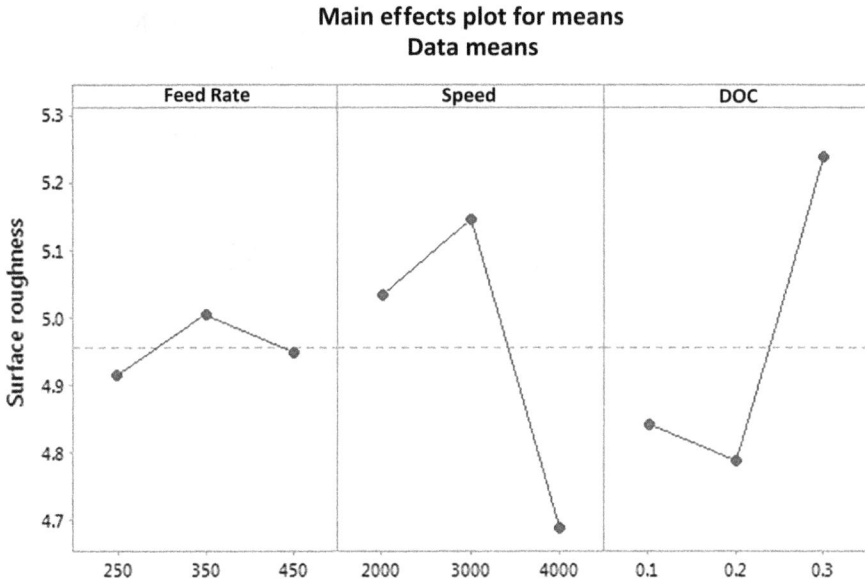

FIGURE 1.4 Mean data analysis for SR.

TABLE 1.6
Optimal Solution for Surface Roughness (SR)

Response	Feed	Speed	DoC	Org	Prd
RA	250	4000	0.2	3.95	4.41

The experiment for optimal solution was performed on a milling machine to verify the experimental and predicted cutting time values. The error among the results is 11%, which is in the range.

1.3.5 CONTOUR PLOT ANALYSIS FOR CUTTING TIME AND SURFACE ROUGHNESS

The contour plot is a plot made among two factors and one selective response variable to find the relation among the factors and the response variable. In this section, contour plots were generated for both the response variables. The contour plots for cutting time were developed using Minitab software.

The results indicate that if the feed rate is increased at a constant value of speed, the cutting time increased. This shows that the feed rate can improve if speed is lower. The region for minimum cutting time is found at highest DoC and highest feed rate levels as seen in figure. The minimum cutting time is found at highest value of DoC and lowest value of speed, and the outcome is shown in figure.

1.3.6 INTERACTION PLOT FOR CUTTING TIME AND SURFACE ROUGHNESS

The interaction plot is a special plot made among two factors and one selective response variable to find the relation among the factors and the response variable. The interaction plot for cutting time and surface roughness is shown in Figures 1.5 and 1.6, respectively. As shown in Figure 1.5, the interaction plot is made among two factors, i.e., speed and feed rate, among which the feed rate is the holding factor which is treated as the constant factor. The cutting time is taken in the y-axis, as shown in figure (a). If the holding factor feed rate is kept fixed and speed is varied, then the cutting time decreases and then increases as the speed increases. The same profile of other figures shown in (b)–(f) is also discussed here. Some profiles linearly decrease, but some profiles of cutting time are zigzag in nature, as shown in (a) and (f). The reason behind these profiles is the nonlinear real-time issue during machining on CNC milling as tool wear rate is a nonlinear parameter. As seen in Figure 1.6 (a), the x-axis factor is speed and the holding factor is feed rate. In figure (a), if the speed is increased, then surface roughness profile decreased for holding values of 250 mm/min and 450 mm/min of feed rate, but for 350 mm/min, the surface roughness is increased and at a certain level, it starts to decrease.

1.3.7 ADEQUACY CHECK ANALYSIS

Three different tests, viz. sequential model sum of squares, lack-of-fit tests, and model summary statistics, are performed to analyze the results. The sequential model sum of squares test in each table shows how the terms of increasing complexity contribute to the model. Residuals possess constant variance as they are scattered randomly around zero in residuals versus the fitted values. Since residuals exhibit no

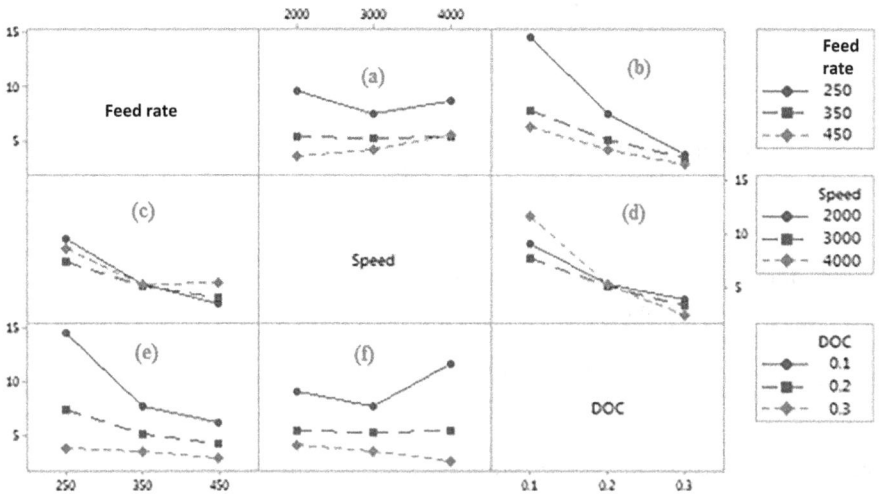

FIGURE 1.5 Interaction plot for CT.

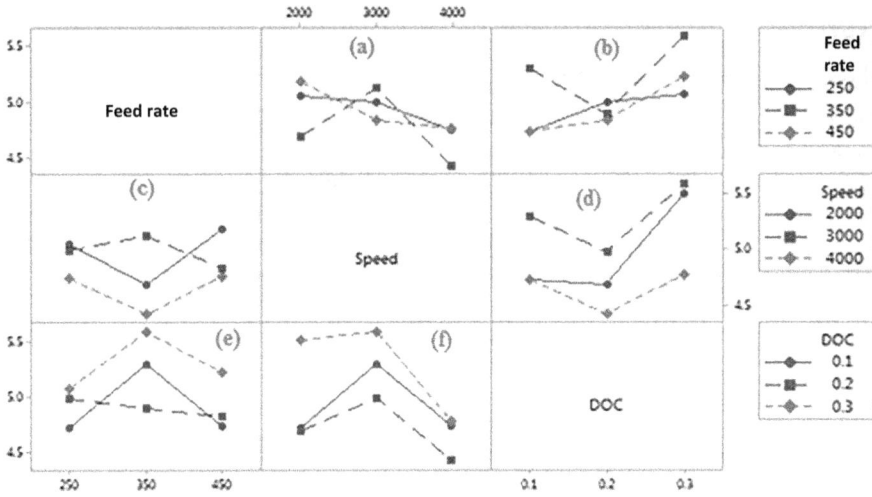

FIGURE 1.6 Interaction plot for SR.

clear pattern, there is no error due to time or data collection order. The intervals at 95% confidence level are used in the present experimental work. Tables 1.7 and 1.8 show the ANOVA results for cutting time and surface roughness, respectively.

Figures 1.7 and 1.8 show the residual plots for cutting time and surface roughness, respectively.

1.3.8 REGRESSION MODELING EQUATION

After performing the adequacy check for both the response variables, final regression modeling equations are generated for CT and surface roughness variables.

TABLE 1.7
ANOVA for Linear Model for CT

Source	DF	Seq. SS	Contribution (%)	Adj. SS	Adj. MS	F-value	p-value
Model	3	159.011	72.30	159.011	53.004	13.92	0
Linear	3	159.011	72.30	159.011	53.004	13.92	0
Feed rate	1	49.195	22.37	49.195	49.195	12.92	0.002
Speed	1	0.454	0.21	0.454	0.454	0.12	0.734
DoC	1	109.362	49.73	109.362	109.362	28.73	0
Error	16	60.907	27.70	60.907	3.807		
Lack-of-fit	11	57.854	26.31	57.854	5.259	8.61	0.014
Pure error	5	3.053	1.39	3.053	0.611		
Total	19	219.918	100.00				

TABLE 1.8
ANOVA for Linear Model for SR

Source	DF	Seq. SS	Contribution (%)	Adj. SS	Adj. MS	F-value	p-value
Model	3	0.69389	33.31	0.69389	0.231297	2.66	0.083
Linear	3	0.69389	33.31	0.69389	0.231297	2.66	0.083
Feed rate	1	0.00256	0.12	0.00256	0.00256	0.03	0.866
Speed	1	0.29929	14.37	0.29929	0.29929	3.45	0.082
DoC	1	0.39204	18.82	0.39204	0.39204	4.52	0.05
Error	16	1.3892	66.69	1.3892	0.086825		
Lack-of-fit	11	1.21127	58.15	1.21127	0.110116	3.09	0.111
Pure error	5	0.17793	8.54	0.17793	0.035587		
Total	19	2.08309	100.00				

Residual plots for cutting time

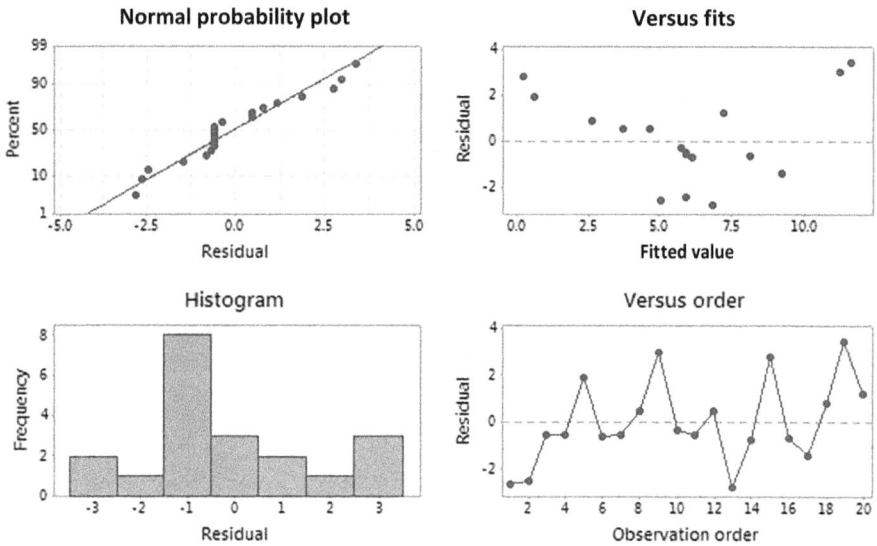

FIGURE 1.7 Residual plot analysis for CT.

$$\text{Cuttingtime} = 49.24 - 0.1410 \text{ feed rate} - 0.00237 \text{ speed} - 91.1 \text{ DoC}$$
$$+ 0.000085 \text{ feed rate} \times \text{feed rate} + 0.000000 \text{ speed} \times \text{speed}$$
$$+ 61.6 \text{ DoC} \times \text{DoC} + 0.000007 \text{ feed rate} \times \text{speed} + 0.1863 \text{ feed rate}$$
$$\times \text{DoC} - 0.01060 \text{ speed} \times \text{DoC} \tag{1.4}$$

Residual plots for Ra – linear

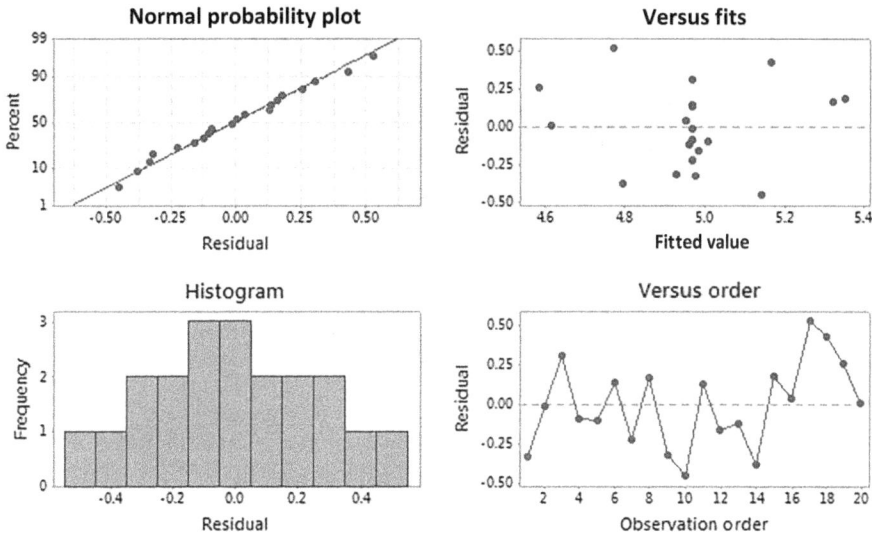

FIGURE 1.8   Residual plot analysis for SR.

$$Ra = 0.72 + 0.00629 \text{ feed rate} + 0.002945 \text{ speed} - 11.96 \text{ DoC} - 0.000009 \text{ feed rate}$$

$$\times \text{ feed rate} - 0.000000 \text{ speed} \times \text{ speed} + 45.45 \text{ DoC} \times \text{ DoC} - 0.000000 \text{ feed rate}$$

$$\times \text{ speed} + 0.00375 \text{ feed rate} \times \text{ DoC} - 0.001850 \text{ speed} \times \text{ DoC} \qquad (1.5)$$

1.3.9   MOGA Optimization Technique

In this section, MOGA optimization technique is applied to regression modeling equations generated by nonlinear technique for responses. Figure 1.9 presents the Pareto analysis results using MOGA approach. Function 1 represents cutting time; function 2 represents surface roughness response equation. After fitness function generation, data were entered into Global Optimization Toolbox, which is available in MATLAB software. Solver was selected as per requirements; for the present study, MOGA is selected as the solver. All the required data were entered in the graphical user interface (GUI) window for the present work. Population size was selected in MATLAB, and after this selection of population size, solver was run and results were generated and exported to workspace window of MATLAB software (Table 1.9). The function values and decision variables by using the MOGA approach are shown in Figure 1.9. Table 1.10 shows the comparison of the results by different techniques.

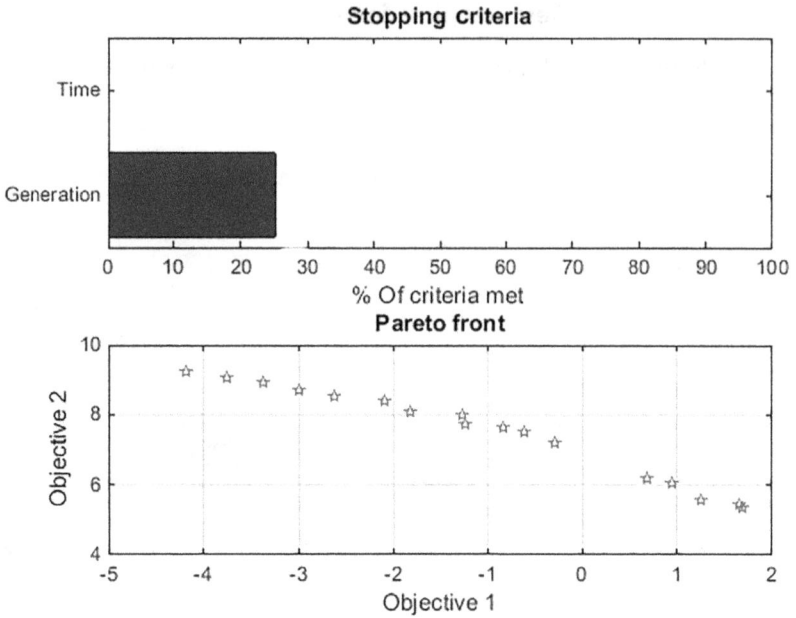

FIGURE 1.9   Pareto analysis results using MOGA.

TABLE 1.9
Function Values and Decision Variables

Experiment No.	CT-($f1$)	SR-($f2$)	Feed ($\times(1)$)	Speed ($\times(2)$)	DoC ($\times(3)$)
1	3.89	6.59	364.9	2000	0.16
2	9.03	6.43	250.1	2000	0.16
3	2.86	6.59	396.8	2000	0.16
4	0.67	9.47	361.7	2882	0.29
5	3.11	11.67	371.3	3817	0.28
6	8.37	6.46	262.6	2000	0.16
7	1.94	7.17	370.1	2048	0.26
8	6.99	6.51	290.1	2000	0.16
9	2.65	11.06	363.4	3515	0.29
10	0.05	8.96	368.0	2705	0.28
11	1.39	7.80	368.6	2342	0.25
12	7.52	6.49	277.2	2001	0.16
13	5.98	6.55	310.8	2003	0.16
14	2.67	6.79	363.4	2004	0.22
15	0.83	8.45	370.0	2619	0.24
16	0.37	8.62	368.0	2587	0.27
17	3.60	11.82	364.5	3824	0.29
18	1.79	10.80	367.1	3516	0.26

TABLE 1.10
Comparison of Results

Technique	Feed	Speed	DOC	CT	RA
Optimal-CT	450	3000	0.30	1.46	-
Optimal-RA	250	4000	0.20	-	4.41
MOGA	400	2000	0.16	2.8	6.5
CD function	437	4000	0.24	3.8	4.4

TABLE 1.11
Predicted and Experimental Values

Response	Optimal Set	Predicted Optimal Value	Actual Value
Cutting time	A450-B-3000C-0.3	1.46	1.75
Surface roughness	A250-B-4000C-0.2	4.41	3.95

1.4 CONCLUSION

In the present study, the effect of process parameters on response parameters of CNC milling machining process was investigated. The feed rate, speed, and depth of cut process were selected with three levels each. The cutting time and surface roughness were investigated in the present study. The following conclusions have been drawn:

1. Using mean data analysis, the rank of each factor for cutting time and surface roughness has been analyzed. The speed is found to be the most significant parameter for surface roughness, and the depth of cut is found to be the most significant parameter for cutting time. The regression equation has also been generated for both the response variables.
2. The optimal solution for each factor was determined and discussed for showing the quality of results. The predicted and experimental results are shown in Table 1.11. The experimental values are in good agreement with the predicted values
3. The optimization of model equations was performed for cutting time and surface roughness using MOGA technique. It was useful to predict the role of optimum solution for milling machining process. The Pareto front results are obtained by using the MATLAB tool.
4. In the present study, mathematical modeling and optimization of process parameters were made for cutting time and surface roughness. This work can be extended to other response variables, i.e., tool wear rate, dimensional deviation, overcut, etc. Also, other parameters can be used to have a more insight into the process.

REFERENCES

1. Oktem, H, Erzurumlu, T & Col, M, 2006, A study of the Taguchi optimization method for surface roughness in finish milling of mold surfaces, *International Journal of Advanced Manufacturing Technology*, vol. 28, no. 7–8, 694–700.
2. Del Prete, A, De Vitis, AA & Anglani, A, 2010, Roughness improvement in machining operations through coupled metamodel and genetic algorithms technique, *International Journal of Material Forming*, vol. 3, no. 1, 467–470.
3. Alam, S, NurulAmin, AKM, Patwari, AU & Konneh, M, 2010, Prediction and investigation of surface response in high speed end milling of Ti-6Al-4V and optimization by genetic algorithm, *Advanced Materials Research*, vol. 83–86, 1009–1015.
4. Chandrasekaran, M, Muralidhar, M, Krishna, CM & Dixit, US, 2010, Application of soft computing techniques in machining performance prediction and optimization: a literature review, *International Journal of Advanced Manufacturing Technology*, vol. 46, no. 5–8, 445–464.
5. Turkes, E, Orak, S, Neseli, S & Yaldiz, S, 2012, Decomposition of process damping ratios and verification of process damping model for chatter vibration, *Measurement*, vol. 45, no. 6, 1380–1386.
6. Kadirgama, K, Noor, MM & Rahman, MM, 2012, Optimization of surface roughness in end milling using potential support vector machine, *Arabian Journal for Science and Engineering*, vol. 37, no. 8, 2269–2275.
7. Patel, KP, 2012, Experimental analysis on surface roughness of CNC end milling process using Taguchi design method, *International Journal of Engineering Science and Technology*, vol. 4, no. 2, 540–545.
8. Pandey, A, Goyal, A & Meghvanshi, R, 2017, Experimental investigation and optimization of machining parameters of aerospace material using Taguchi's DOE approach, *Materials Today: Proceedings*, vol. 4, no. 8, 7246–7251.
9. Pinar, AM, 2013, Optimization of process parameters with minimum surface roughness in the pocket machining of AA5083 aluminum alloy via Taguchi method, *Arabian Journal for Science and Engineering*, vol. 38, no. 3, 705–714.
10. Goyal, A., Pandey A., Sharma P., and Sharma SK, 2017, Study on Ni-based super alloyusing cryogenic treated electrode by Taguchi methodology, *Materials Today: Proceedings*, vol. 4, no. 2, 2068–2076.

2 Multi-Objective Optimization for Improving Performance Characteristics of Novel Curved EDM Process Using Jaya Algorithm

Diwesh B. Meshram
Central Institute of Plastics Engineering and Technology

Yogesh M. Puri
Visvesvaraya National Institute of Technology

Neelesh Kumar Sahu
Medi-Caps University

CONTENTS

2.1 INTRODUCTION

This era marks an ever more so increasing demand for new and innovative products, which further face a fiercely competitive market. The existing products require modifications in design and a continuous improvement in performance for sustainable growth. Some customized products require a high accuracy with complex machining geometry. It is achieved by non-conventional machining processes compared to conventional machining. Electrical discharge machining (EDM) process is the best solution for achieving products with desired requirements.

The die-sinking EDM process is based on the principle of spark erosion. The removal of workpiece material from the customized electrode is performed by a series of electric sparks inside the dielectric medium [1].Hard workpiece material is possible to machine through the EDM process It is possible to machine hard workpiece materials using the EDM process. Simplified electrode designs and manufacturing concepts have been discussed to improve the performance of the EDM [2]. The dielectrics (liquid and gas) act as a medium for increasing the electric sparks and absorbing the heat generated during machining [3,4].

The year 2002 marks the start of the era of machining curved channels. An L-shaped hole was created by connecting two different lines perpendicular to each other with a customized mechanism [5]. The replacement of the straight channel with a U-shaped channel using an electrode curved motion generator was discussed in 2008 [6]. In 2010, the automatic discharge gap controller (ADGC) mechanism was developed with a microrobot for machining long curved holes [7]. The hybrid mechanism was used for machining conductive hard materials. Various methods for machining curved channels were analysed and discussed in the year 2013 [8]. After two years, the importance of the curved channel was discussed and a mechanism based on the suspended ball electrode was developed [9]. In the next year, a modification was suggested for machining helical channels and its versatility was discussed [10]. A customized trajectory for machining profiles of different shapes, such as S-shaped and J-shaped profiles, was made possible on the EDM machine with certain curvature [11]. In 2018, various mechanisms available for machining curved channels using EDM were reviewed [12].

In sinker EDM, the linear ram head is coupled with the electrode and moves up and down on the stationary workpiece. This process is called linear machining. This process finds its application in cases where certain industries require a customized curvature trajectory to be machined on the workpiece, such as cooling channels in mould. This trajectory may be machined on the external and internal surfaces of the workpiece material. It is possible to machine the external curvature trajectory irrespective of the internal surface. Due to limitations in machining the internal channel using the sinker EDM, a novel curved EDM machining mechanism is proposed. This mechanism was integrated with mechatronics systems for achieving the curvature trajectory. An electronic control system (ECS) was designed and programmed for the synchronization of the control system of the sinker EDM machine. The linear motion of the defined electrode is changed to an oscillating motion by using an electrode-locating mechanism. The block diagram of the proposed mechanism is shown in Figure 2.1.

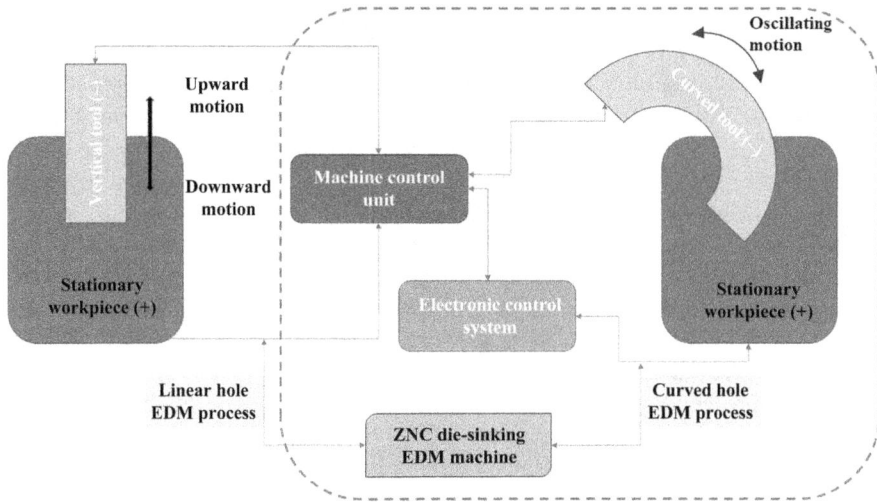

FIGURE 2.1   Block diagram of the novel curved EDM process.

In view of the huge demand for high-precision machining techniques employable in the development of complex channels and holes, requisite for moulding applications, especially in the case of cooling systems of plastic moulds, our work emphasizes an experiment conducted for machining curved channels using the novel curved EDM process. The development of the curvature trajectory and its optimization is based on eight input machining parameters and two output process characteristics. Taguchi orthogonal array (L12) is obtained using Minitab statistical software based on the input variables [13,14]. The results obtained for MRR and EWR are evaluated and are effectively analysed by analysis of variance (ANOVA) [15–17]. Further, regression analysis is adopted for obtaining the interrelationship between the input and output variables [18–20]. The regression equation obtained acts as an objective function for the multi-objective optimization. The advanced Jaya algorithm is implemented for achieving the optimal solution for the output characteristics [21–24].

2.2   EXPERIMENTAL METHODOLOGY

The experimental methodology comprises of an overview of the novel curved EDM mechanism, statistical analysis by ANOVA, multi-objective optimization by regression analysis, and Jaya algorithm.

2.2.1   DESIGN, DEVELOPMENT AND OPERATION OF THE NOVEL CURVED EDM MECHANISM

This mechanism is used for internal and external machining of curved channels on the workpiece material. The manufacturing of curved electrodes is an important application of the mechanism, wherein a programmable ECS with mechanical and electrical components is selected and mapped with the z-axis numerical control (ZNC) die-sinking EDM machine.

The curved machining methodology is transformed from conceptual stage to design stage by modelling the individual components and assembling them in Siemens (NX 10) software. Further, the individual component is developed as per the design. The sequential assembly of actual components is performed, and the complete mechanism is installed on the EDM machine, as shown in Figure 2.2.

The mechanism consists of mechanical and electrical components, viz. the major components such as a permanent magnet direct control (PMDC) motor, two L-shaped flanges, a timing belt, two timing gears, a bearing support assembly, an acrylic transparent tank, aluminium supporting blocks, nylon hexagonal bolts and steel nuts, a collet, a curved electrode with electrode supporting mechanism (ESM) and the workpiece. A programmable ECS is developed for controlling the desired functions of the interlinked components.

The operation of novel curved machining mechanism involves the motion transmission to the curved electrode through various linked components. A customized sensor is provided for sensing the linear motion of the ram head. The signals are transferred to the PMDC motor via ECS to obtain forward–reverse motion with designated rpm. The small and big gears are mounted on the shaft of the motor and bearing shaft, respectively. The timing belt connects and transmits the motion to the bearing support assembly placed inside the transparent acrylic tank. A hollow aluminium support is used for adjusting the height of the curved electrode with the workpiece. Customized nylon bolts are manufactured, acting as an insulator between the machine tank and components. The collet is attached at the end of the bearing shaft for holding the ESM. Finally, transmission of motion from various components is provided to the curved electrode in the forward–reverse direction. The spark is initiated on the workpiece based on the spark gap. The operating machining parameters are decided and selected for machining curved channels through the mechanism.

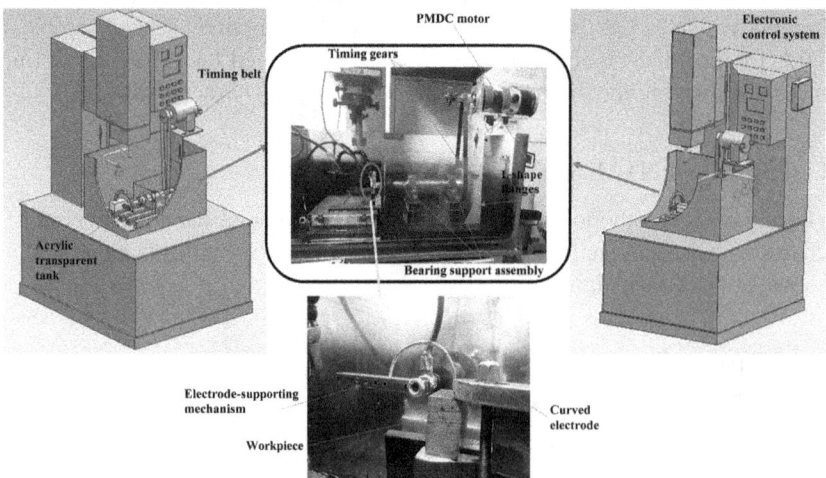

FIGURE 2.2 Actual mechanism of the novel curved EDM process.

2.2.2 Experimental Investigation of Curved Machining Mechanism

The mechanism is unique and based on the movement of curvature trajectory in a specific direction. The selection of curved electrode and workpiece material was based on the pilot study executed through this mechanism. The selected workpiece material is mostly used in core and cavity plates of plastic injection mould. The necessary cooling channel was machined inside the mould base for improving the quality of part and reducing cycle time. The details of the curved electrode and workpiece material are shown in Table 2.1.

The actual control variables with machining responses were identified, and experiments were formulated. The decision values of the control variables were estimated on the basis of the minimum and maximum operating range. Pulse ON time, duty cycle, sparking current, bi-pulse current, sparking time, lifting time, gap voltage and sensitivity are the control variables used in machining the curvature channel, as shown in Table 2.2. The two major machining responses are estimated and evaluated, i.e. electrode wear rate (EWR) and material removal rate (MRR), for performance improvement in the novel curved EDM process [25].

TABLE 2.1
Technical Specifications with Curved Electrode and Workpiece Material Chemical Composition

	Curved Electrode	Workpiece
Material with chemical composition (%)	Pure copper (99% Cu)	Oil hardening non-shrinkage steel (OHNS) C (carbon) – 0.94, Mn (magnesium) – 0.99, W (tungsten) – 0.49, Cr (chromium) – 0.50, V (vanadium) – 0.99, Fe (iron) – remaining
Technical specification	Diameter – 33 mm Cross-sectional thickness: 3.2 mm -	Length – 40 mm, Width – 15 mm Height – 40 mm

TABLE 2.2
Control Variables with the Defined Levels

			Level 1	Level 2
Notation	Control Variables (Units)		Maximum Value	Minimum Value
A	TON	Pulse ON time (µs)	0.5	1,050
B	D	Duty cycle (µs)	5	181
C	IP	Sparking current (A)	1	35
D	IB	Bi-pulse current (A)	1	3
E	SPK	Sparking time (µs)	1	20
F	LFT	Lifting time (µs)	1	2
G	GAP	Gap voltage (V)	1	99
H	SEN	Sensitivity (mm/min)	1	100

The optimum machining parametric combinations of the machining variables were obtained by Taguchi orthogonal array (L12). Minitab 17 statistical software was used for obtaining the required outputs based on the inputs [26]. It also provided the statistics related to graphical analysis and mathematical modelling.

The MRR and EWR were calculated using the following formulae:

$$EWR\left(mm^3/min\right)$$

$$= \frac{\left(\text{Weight of the electrode}\left(\text{before}\right)\right) - \left(\text{Weight of the electrode}\left(\text{after}\right)\right)}{\left(\left(\text{Density of electrode material}\right) \times \left(\text{Machining time}\right)\right)} \quad (2.1)$$

$$MRR\left(mm^3/min\right)$$

$$= \frac{\left(\text{Weight of the Workpiece}\left(\text{before}\right)\right) - \left(\text{Weight of the Workpiece}\left(\text{after}\right)\right)}{\left(\left(\text{Density of electrode material}\right) \times \left(\text{Machining time}\right)\right)} \quad (2.2)$$

The accurate weights of the electrode and workpiece material were measured before and after machining using a high-precision weighing balance machine (model: SB-24001DR). The density of copper electrode and OHNS workpiece material is required for estimating the volumetric rate of machining responses, and the values are 8.96 and 8.67 g/cm³, respectively.

The machining responses were evaluated based on the experiments performed. The values of EWR and MRR are presented in Table 2.3.

The values of machining responses were measured thrice for reducing the error in measurements. Based on the input control variables, the performance was analysed.

TABLE 2.3
Calculated Experimental Values for EWR and MRR

Exp. No.	A	B	C	D	E	F	G	H	EWR (mm³/min)	MRR (mm³/min)
1	0.5	5	1	1	1	1	1	1	0.05	0.49
2	0.5	5	1	1	1	2	99	100	0.07	0.41
3	0.5	5	35	3	20	1	1	1	0.10	0.56
4	0.5	181	1	3	20	1	99	100	0.02	0.54
5	0.5	181	35	1	20	2	1	100	0.18	0.67
6	0.5	181	35	3	1	2	99	1	0.04	0.59
7	1080	5	35	3	1	1	99	100	0.20	0.69
8	1080	5	35	1	20	2	99	1	0.08	0.51
9	1080	5	1	3	20	2	1	100	0.06	0.57
10	1080	181	35	1	1	1	1	100	0.22	0.62
11	1080	181	1	3	1	2	1	1	0.12	0.44
12	1080	181	1	1	20	1	99	1	0.16	0.51
Average									0.11	0.55

2.2.3 STATISTICAL ANALYSIS FOR THE MACHINING
RESPONSES USING ANALYSIS OF VARIANCE

The statistical analysis was performed on the calculated values based on the Taguchi orthogonal array. Initially, the normality test was adopted for determining the obtained values following normal distribution using a significance level (P-value) of 0.05. The P-values of EWR and MRR are greater than the significant level, stating that the data are normal by Anderson–Darling test, as shown in Figures 2.3 and 2.4.

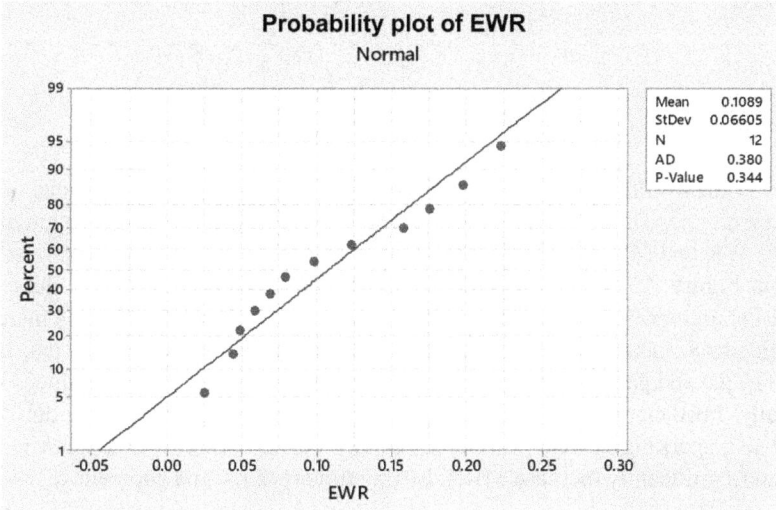

FIGURE 2.3   Normality plot for EWR.

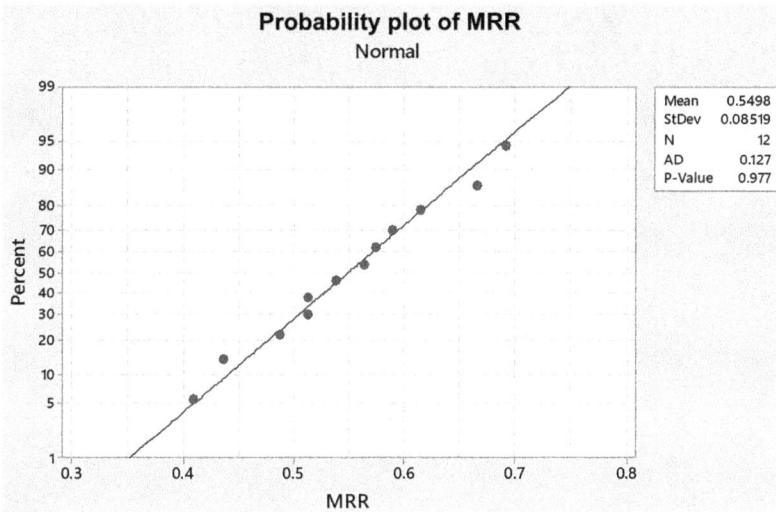

FIGURE 2.4   Normality plot for MRR.

The main effects plot of minimum and maximum levels was analysed for the machining responses. It is based on signal-to-noise (S/N) ratio in which the condition is selected as minimum of EWR and maximum of MRR. The S/N ratio is generally used to find the best setting of control variables that are involved in production process. The governing equations for EWR (smaller is better) and MRR (larger is better) are given by:

$$\eta = -10\log\left[\frac{1}{n}\sum_{i=i}^{n}y_i^2\right] \tag{2.3}$$

$$\eta = -10\log\left[\frac{1}{n}\sum_{i=1}^{n}\left(\frac{1}{y_i^2}\right)\right] \tag{2.4}$$

Here, η is the resultant S/N ratio (units: dB), n is the number of replications of each experiment, and y_i is the process parameters of the ith experiment. The main effects plot of EWR and MRR are plotted in Figures 2.5 and 2.6.

From Figure 2.5, it is observed that pulse ON time (0.5–1,080 µs), duty cycle (5–181 µs) and sparking current (1A-35A) reduce EWR. Further, EWR is increased by the control variables bi-pulse current (1A-3A), sparking time (1–20 µs), lifting time (1–2 µs) and gap voltage (1 V–99 V). Sensitivity (1–100mm/min) reduces EWR gradually. Further, the control variables pulse ON time (0.5–1,080 µs), duty cycle (5–181 µs), sparking current (1A-35A), bi-pulse current (1A-3A) and sparking time (1–20 µs) significantly increase MRR. Lifting time (1–2 µs) and gap voltage (1–99 V)

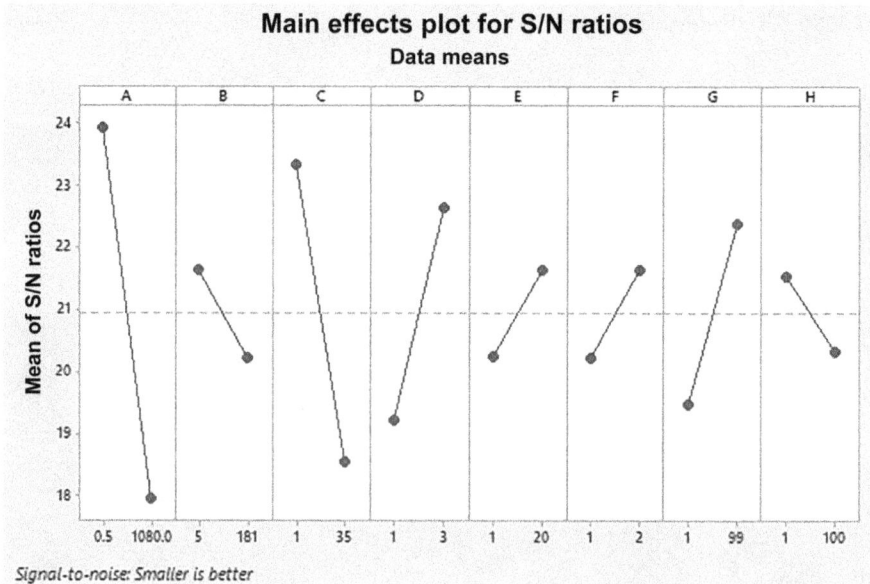

Main effects plot for S/N ratios
Data means

Signal-to-noise: Smaller is better

FIGURE 2.5   Main effects plot for EWR.

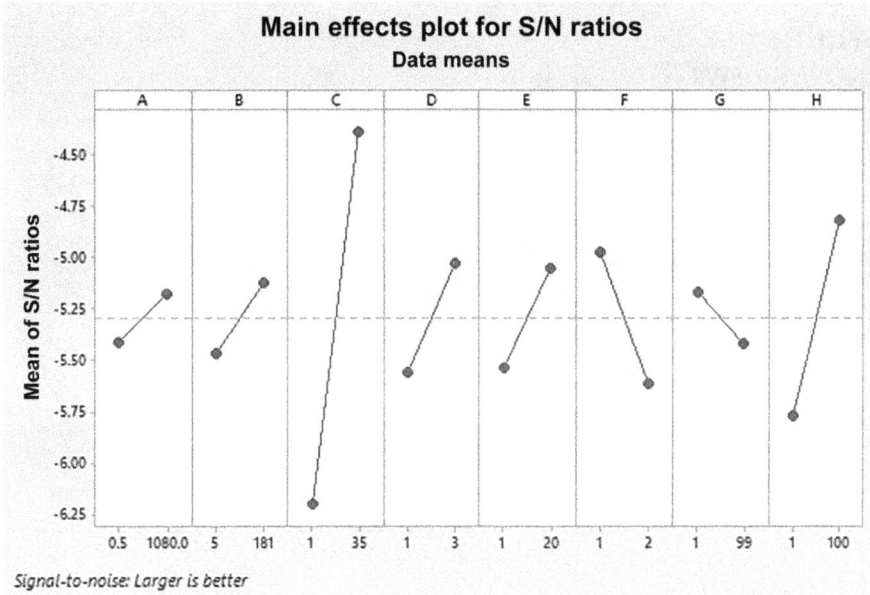

FIGURE 2.6   Main effects plot for MRR.

reduce MRR. Sensitivity (1–100 mm/min) significantly increases MRR, as shown in Figure 2.6.

In order to determine the characteristics and the relative importance of control variables, analysis of variance (ANOVA) was performed. ANOVA results were estimated by separating the total variability into contributions by each of the control variables and error. ANOVA results for EWR and MRR are shown in Tables 2.4 and 2.5. The general linear model (GLM) in ANOVA was adopted for identifying the individual contribution from the total contribution of machining responses. The additional effect of interaction of control variable was determined and analysed in the GLM. Further, the forward selection method was adopted in Minitab 17 statistical software for fitting the values of machining responses with the addition of most significant control variables.

The major contributor was found to be pulse ON time (25.01%), which led to the minimization of EWR. Further, the sparking current (19.47%) also contributed to reducing the EWR. The interaction of bi-pulse current and gap voltage (10.47%) also significantly contributed to minimizing the EWR. The major control variables significantly affecting the EWR based on the p-value are sparking current, sensitivity and the interaction of sparking current and sensitivity.

The results of ANOVA estimated for MRR are presented in Table 2.5. It is found that the major contributing factor that leads to maximizing the MRR when using OHNS workpiece is the sparking current. The interaction of pulse ON time and duty cycle also has a significant contribution to achieving the maximum MRR. From the p-value of ANOVA estimation as shown in Table 2.5 of MRR, the factors that have maximum and minimum effect on MRR are sparking current and lifting time, respectively.

TABLE 2.4
ANOVA for EWR

Notation	DF	Seq. SS	Adj. SS	F-Value	P-Value	Contribution (%)
A	1	0.011999	0.000149	1.8	0.408	25.01
B	1	0.003199	0.000091	1.09	0.486	6.67
C	1	0.009342	0.009191	110.68	0.06	19.47
D	1	0.003531	0.000700	8.43	0.211	7.36
G	1	0.002035	0.000474	5.71	0.252	4.24
H	1	0.003199	0.008041	96.83	0.064	6.67
AD	1	0.001706	0.001723	20.75	0.138	3.56
BC	1	0.003189	0.000468	5.63	0.254	6.65
CH	1	0.004678	0.009673	116.49	0.059	9.75
DG	1	0.005023	0.005023	60.49	0.081	10.47
Error	1	0.000083	0.000083	-	-	0.17
Total	11	0.047985	-	-	-	100.00

TABLE 2.5
ANOVA for MRR

Notation	DF	Seq. SS	Adj. SS	F-Value	P-Value	Contribution (%)
A	1	0.00063	0.00008	0.8	0.535	0.79
B	1	0.00116	0.00082	7.49	0.223	1.45
C	1	0.03874	0.02375	216.93	0.043	48.52
E	1	0.0016	0.00000	0.05	0.856	2.00
F	1	0.00405	0.01156	105.6	0.062	5.07
G	1	0.00063	0.00022	1.96	0.395	0.79
AB	1	0.02139	0.00044	4	0.295	26.79
BC	1	0.00269	0.00695	63.48	0.079	3.37
BE	1	0.00271	0.00063	5.76	0.251	3.39
FG	1	0.00613	0.00613	56.01	0.085	7.68
Error	1	0.00011	0.00011	-	-	0.14
Total	11	0.07984	-	-	-	100.00

2.2.4 MULTI-OBJECTIVE OPTIMIZATION FOR THE OPTIMUM MACHINING RESPONSES

On the basis of the experimental study and the implementation of statistical investigation, it was found that it is difficult to obtain the interrelationship between the eight control variables for optimizing the two machining responses; hence, multiple regression analysis is employed. The optimum values are validated using a nature-based Jaya multi-objective optimization algorithm.

2.2.5 MULTIPLE REGRESSION ANALYSIS

Multiple regression analysis was implemented for the prediction of optimum values of minimum EWR and maximum MRR based on the control variables. The relationship between the control variables and machining responses was obtained through the regression equation. The general linear model (GLM) was used with the two-way interaction of control variables for obtaining the regression equation for EWR [27]. The forward selection method was selected in regression type with $\alpha = 0.05$. The continuous predictor standardization levels were coded with -1 and $+1$ for minimum and maximum values of control variables. The regression equations for EWR (Eq. 2.5) and MRR (Eq. 2.6) were obtained and are presented as follows:

$$EWR = -0.0292 + 0.000084(A) - 0.000186(B) + 0.003501(C) - 0.0830(D)$$

$$+ 0.001953(G) - 0.000988(H) - 0.000047(AD) + 0.000008(BC)$$

$$+ 0.000095(CH) - 0.000886(DG) \tag{2.5}$$

$$MRR = 0.4353 + 0.000017(A) - 0.000322(B) + 0.001285(C) + 0.001801(E)$$

$$+ 0.0445(F) + 0.004054(G) - 0.000000(AB) + 0.000030(BC)$$

$$- 0.000018(BE) - 0.002589(FG) \tag{2.6}$$

The regression analysis model prediction was based on two prediction factors, i.e. coefficient of determination (R^2) and adjusted R^2 (R^2_{adj}). The R-square explains how the values of machining responses are fit in a curve or line with respect to the control variables. Further, the adjusted R-square adjusts the values of the machining responses with respect to the control variables (Table 2.6).

TABLE 2.6
Coefficients of Regression for EWR

Notation	Coeff.	SE Coeff.	95% CI	T-Value	P-Value	VIF
Constant	0.108920	0.002630	(0.07550, 0.14235)	41.41	0.015	-
A	−0.005580	0.004160	(−0.05843, 0.04727)	−1.34	0.408	2.5
B	−0.004240	0.004050	(−0.05575, 0.04727)	−1.05	0.486	2.37
C	0.033900	0.003220	(−0.00704, 0.07483)	10.52	0.06	1.5
D	0.013230	0.004560	(−0.04467, 0.07112)	2.9	0.211	3
G	0.008890	0.003720	(−0.03838, 0.05616)	2.39	0.252	2
H	0.035450	0.003600	(−0.01032, 0.08121)	9.84	0.064	1.87
AD	−0.025420	0.005580	(−0.09633, 0.04548)	−4.56	0.138	4.5
BC	0.011470	0.004830	(−0.04993, 0.07288)	2.37	0.254	3.38
CH	0.079680	0.007380	(−0.01412, 0.17347)	10.79	0.059	7.87
DG	−0.043400	0.005580	(−0.11431, 0.02750)	−7.78	0.081	4.5
Model summary		S	R^2	$R^2_{(adj)}$	-	-
		0.009112	99.83%	98.10%	-	-

TABLE 2.7
Coefficient of Regression for MRR

Notation	Coeff.	SE Coeff.	95% CI	T-Value	P-Value	VIF
Constant	0.54979	0.00302	(0.51141, 0.58817)	182.01	0.003	-
A	−0.00427	0.00478	(−0.06496, 0.05641)	−0.89	0.54	2.5
B	−0.01132	0.00414	(−0.06388, 0.04124)	−2.74	0.22	1.88
C	0.06856	0.00466	(0.00941, 0.12771)	14.73	0.04	2.37
E	0.00085	0.0037	(−0.04615, 0.04786)	0.23	0.86	1.5
F	−0.04251	0.00414	(−0.09506, 0.01005)	−10.28	0.06	1.88
G	0.00833	0.00595	(−0.06722, 0.08388)	1.4	0.4	3.88
AB	−0.01282	0.00641	(−0.09423, 0.06860)	−2	0.3	4.5
BC	0.04421	0.00555	(−0.02630, 0.11472)	7.97	0.08	3.38
BE	−0.01538	0.00641	(−0.09680, 0.06604)	−2.4	0.25	4.5
FG	−0.06344	0.00848	(−0.17114, 0.04427)	−7.48	0.09	7.88
Model summary		S	R^2	$R^2_{(adj)}$	-	
		0.0104639	99.86%	98.49%		

The values of R^2 and $R^2_{(adj)}$ for EWR and MRR are found to be 99.83% and 98.10% and 99.86% and 98.49%, respectively. The selected regression model for optimizing EWR and MRR are justified and validated by R^2 and $R^2_{(adj)}$. It is also found that the contribution of sparking time, bi-pulse current and sensitivity to EWR and MRR is negligible and hence can be eliminated. The values of control variables along with their levels are well fitted by using $R^2_{(adj)}$. All the predicted values are well fitted in the regression model.

The predicted regression model identifies the major contributors to the performance improvement in EWR and MRR. From Table 2.7, it can be seen that pulse ON time (25.01%) and sparking current (19.47%) are the major control variables of EWR. The interaction of pulse ON time and duty cycle (20.58%) significantly contributes to EWR. From the ANOVA results shown in Table 2.8 of MRR, it is concluded that the major control variable is sparking current (48.52%). The significant contribution of the interaction of pulse ON time and duty cycle (26.79%) increases the MRR in the novel curved EDM process (Table 2.9).

2.3 JAYA ALGORITHM

The word Jaya is of Sanskrit origin, which means victory. This algorithm was proposed by Rao [28,29]. It illustrates the principle of optimization in which the obtained results are approaching toward the accepted outcomes and sidesteps rejected outcomes. Unlike other non-traditional algorithms, it needs only basic optimization knowledge such as the number of design variables, population size and objective functions. Figure 2.7 displays the flow chart of the algorithm 'Jaya'. An objective function (electrode wear rate (EWR) or material removal rate (MRR)) needs to be minimized or maximized for each iteration u assuming there are h numbers of input

TABLE 2.8
ANOVA of Regression for EWR

Notation	DF	Seq. SS	Adj. MS	F-Value	P-Value	Contribution (%)
Regression	10	0.047902	0.004790	57.68	0.102	99.83
A	1	0.011999	0.000149	1.80	0.408	25.01
B	1	0.003199	0.000091	1.09	0.486	6.67
C	1	0.009342	0.009191	110.68	0.06	19.47
D	1	0.003531	0.000700	8.43	0.211	7.36
G	1	0.002035	0.000474	5.71	0.252	4.24
H	1	0.003199	0.008041	96.83	0.064	6.67
AD	1	0.001706	0.001723	20.75	0.138	3.56
BC	1	0.003189	0.000468	5.63	0.254	6.65
CH	1	0.004678	0.009673	116.49	0.059	9.75
DG	1	0.005023	0.005023	60.49	0.081	10.47
Error	1	0.000083	0.000083	-	-	0.17
Total	11	0.047985	-	-	-	100.00

TABLE 2.9
ANOVA of Regression for MRR

Notation	DF	Seq. SS	Adj. MS	F-Value	P-Value	Contribution (%)
Regression	10	0.07973	0.007973	72.82	0.091	99.86
A	1	0.00063	0.000088	0.8	0.535	0.79
B	1	0.00116	0.00082	7.49	0.223	1.45
C	1	0.03874	0.023752	216.93	0.043	48.52
E	1	0.0016	0.000006	0.05	0.856	2.00
F	1	0.00405	0.011563	105.6	0.062	5.07
G	1	0.00063	0.000215	1.96	0.395	0.79
AB	1	0.02139	0.000438	4	0.295	26.79
BC	1	0.00269	0.006951	63.48	0.079	3.37
BE	1	0.00271	0.000631	5.76	0.251	3.39
FG	1	0.00613	0.006132	56.01	0.085	7.68
Error	1	0.00011	0.000109	-	-	0.14
Total	11	0.07984	-	-	-	100.00

factors such as $j = 1, 2,..., m$ (pulse ON time, duty cycle, sparking current, bi-pulse current, sparking time, lifting time, gap voltage and sensitivity) and n numbers of function values ($k = 1, 2,..., n$). The best objective function found in the entire population is represented as MRR $_{best}$, and the worst value is represented as MRR $_{worst}$. Further, assuming $E_{j,k,i}$ represents the jth factor for the kth population at uth iteration, this factor can be changed by the following equation:

$$E'_{j,k,i} = E_{j,k,i} + r_{1,j,i}\left[\left(E_{j,best,i}\right) - \left|E_{j,k,i}\right|\right] - r_{2,j,i}\left[\left(E_{j,worst,i}\right) - \left|E_{j,k,i}\right|\right] \quad (2.7)$$

```
┌─────────────────────────────────────────┐
│  Define EDM control parameters (input variables), no. of │
│  points between limits and termination criteria          │
└─────────────────────────────────────────┘
```

```
┌─────────────────────────────────────────┐
│  Find min (best) and max (worst) values of EWR or         │
│  Find max (best) and min (worst) values of MRR            │
└─────────────────────────────────────────┘
```

```
┌─────────────────────────────────────────┐
│  Modify solutions based on worst and best solutions       │
│  E'ⱼₖⱼ = Eⱼₖⱼ + r₁,ⱼⱼ[(Eⱼ,best,ᵢ) − |Eⱼₖ,ᵢ|] − r₂,ⱼⱼ[(Eⱼ,worst,ⱼ) − |Eⱼₖⱼ|] │
└─────────────────────────────────────────┘
```

$$E'_{jk,j} = E_{jk,j} + r_{1,jj}\left[\left(E_{j,best,i}\right) - \left|E_{jk,i}\right|\right] - r_{2,jj}\left[\left(E_{j,worst,j}\right) - \left|E_{jk,j}\right|\right]$$

Yes　　　　　　　⟨ Is solution corresponding to E'ⱼₖ,ᵢ better than that corresponding to Eⱼₖ,ᵢ ? ⟩　　　　　No

```
┌────────────────────┐                      ┌────────────────────┐
│ Accept and replace │                      │ Keep the           │
│ the previous solution │                   │ previous solution  │
└────────────────────┘                      └────────────────────┘
```

No　　　⟨ Is termination criteria satisfied? ⟩　　　Yes

```
┌────────────────────────────┐
│  Report the optimum solution │
└────────────────────────────┘
```

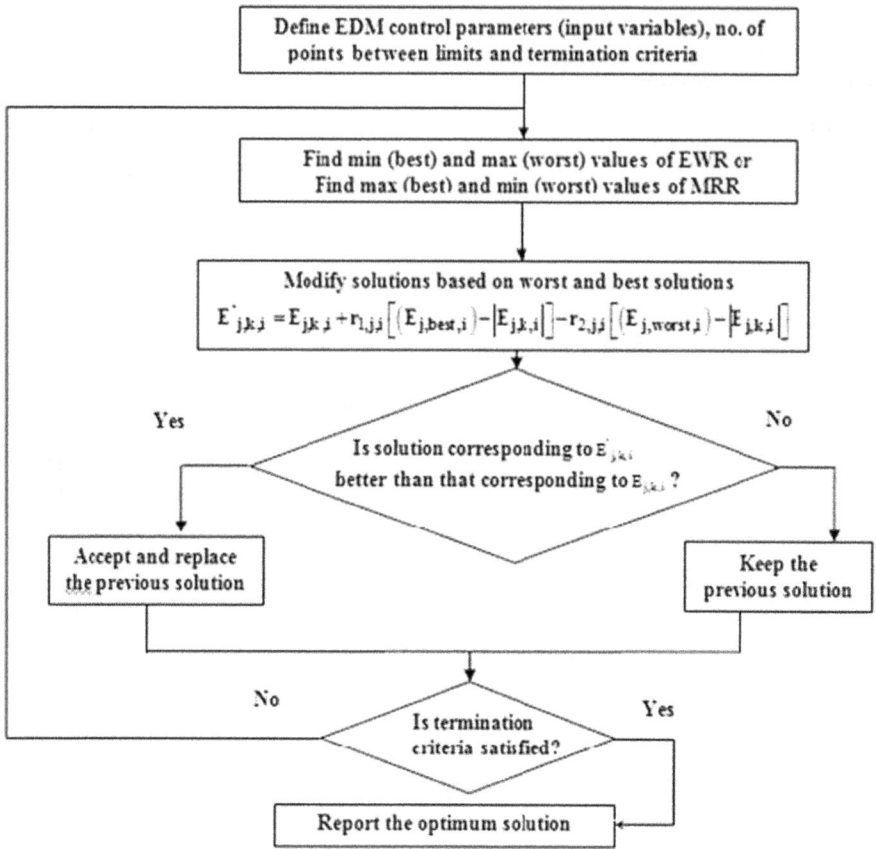

FIGURE 2.7　Flow chart of Jaya algorithm for the novel curved EDM process.

where $E_{j,\,best,\,i}$ represents the point of the EDM parameters j for the best of EWR or MRR value and $E_{j,\,worst,\,i}$ represents the point of the EDM parameters j for the worst of EWR or MRR for ith iteration. $E'_{j,\,k,\,i}$ represents the customized value of $E_{j,\,k,\,i}$, while $r_{1,j,i}$ and $r_{2,j,i}$ are two random numbers for the jth input factor for the ith iteration between 0 and 1. Equation 2.1 is made of two sections. The first section, i.e. $r_{1,j,i}\left[\left(E_{j,\text{best},i}\right) - \left|E_{j,k,i}\right|\right]$, represents curiosity for the best solution, whereas the second section, i.e. $r_{2,j,i}\left[\left(E_{j,\text{worst},i}\right) - \left|E_{j,k,i}\right|\right]$, represents escaping from the worst solution. When $E'_{j,\,k,\,i}$ shows a better value of function, it was agreed; otherwise, that $E_{j,\,k,\,i}$ will be held as it is. For the next iteration, all the accepted function values are considered as input. In the context of input factor, random numbers explore broad search space. $\left|E_{j,k,i}\right|$ gives an absolute value to the solution, improving the algorithm further. This algorithm was successfully demonstrated in the present research for multi-objective optimization of novel curved EDM process.

2.4 RESULTS AND DISCUSSION

The curved channel was machined by the novel curved EDM process. Control variables were selected for improving the performance of the EDM process through the machining responses. Taguchi orthogonal array (L12) was implemented, and experiments were formulated. The experimental values of EWR and MRR were fitted using a normality plot. The optimum values of control variables were used for the prediction and confirmation of performance improvement in the machining responses.

The methodology envisaged the design of a model experiment with a combination of the control variables along with their levels A1B1C1D2E2F2G2H1 and A2B2C2D2E2F1G1H2 to obtain EWR and MRR, respectively, by the main effects plot. The ANOVA results showed that the pulse ON time and sparking current are the major factors affecting the performance with respect to EWR and MRR. Further, it was also observed that the contribution of sparking time and bi-pulse current combination is negligible and is eliminated from the ANOVA table for EWR and MRR.

The relationship between the control variable and machining responses was identified by the multiple regression analysis. The predicted model displayed a good fit using the GLM. The regression values for EWR and MRR were estimated for identifying the variations observed between the experimental values. From the ANOVA table for regression, it was found the impacts of bi-pulse current and lifting time on EWR and bi-pulse current and sensitivity on MRR were negligible and eliminated.

Further, the response optimizer was used for analysing the results obtained from the ANOVA and regression by GLM. From Figure 2.8, the predicted optimum values of EWR and MRR are found to be −0.0174 and 0.763 mm³/min with the desired combination of A1B1C1D2E1F1G2H1, respectively. The interaction plots of MRR and EWR are shown in Figures 2.9 and 2.10.

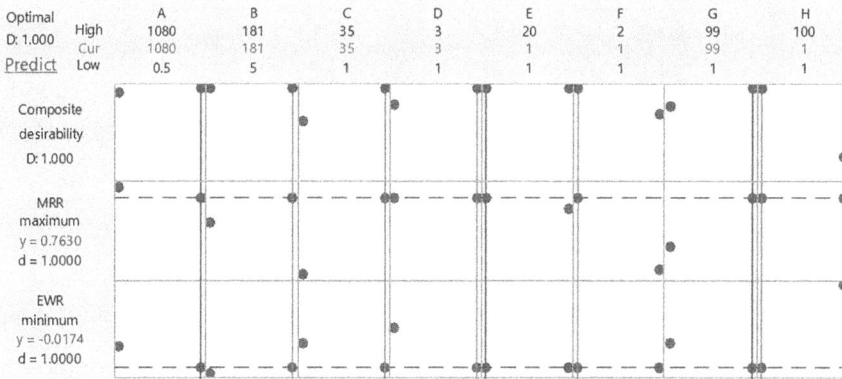

FIGURE 2.8 Response optimizer for EWR and MRR using ANOVA.

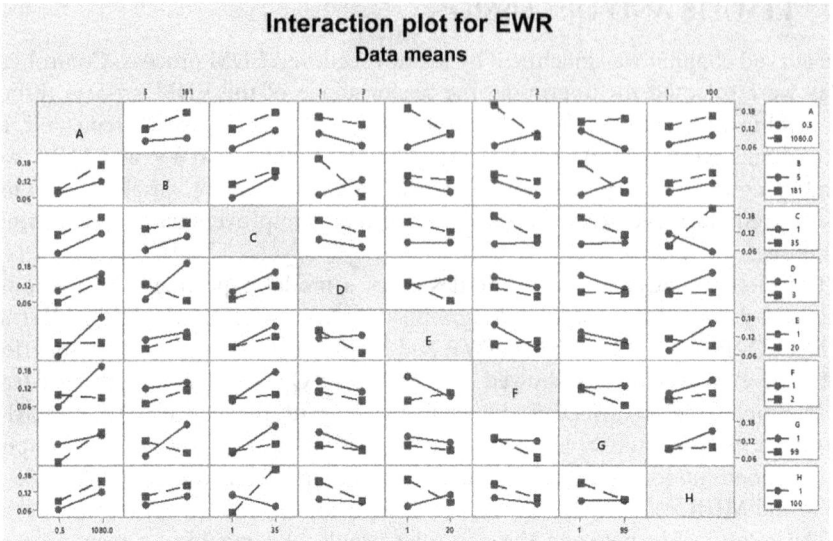

FIGURE 2.9   Interaction plot for EWR obtained from ANOVA.

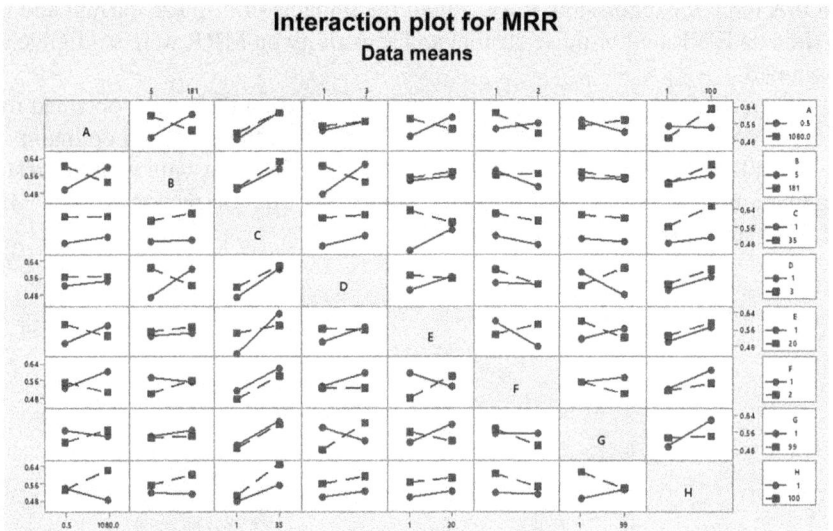

FIGURE 2.10   Interaction plot for MRR obtained from ANOVA.

A comparative analysis of the experimental values and predicted regression values for EWR and MRR is shown in Figure 2.11. It was found that there is less deviation in the results of EWR and MRR.

Further, the multi-objective regression analysis was performed and the optimum results were obtained using the response optimizer through GLM. The optimal

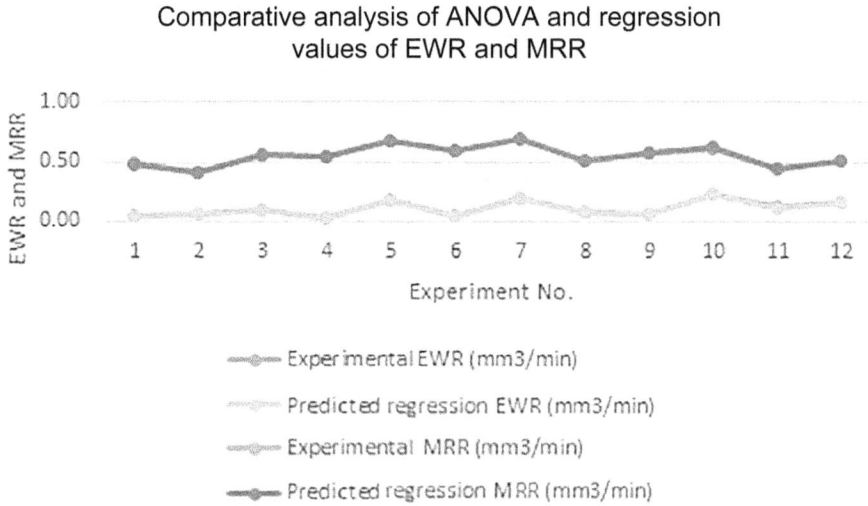

FIGURE 2.11 Comparative analysis of experimental and predicted values of EWR and MRR.

values of EWR and MRR were found to be −0.0174 and 0.7339 mm³/min with the combination of A2B2C2D2E2F1G2H1, as shown in Figure 2.12.

The contour plot was used for analysing the behaviour, corresponding to minimum EWR and maximum MRR. From Figure 2.13, the results are validated towards the minimization of EWR with the optimum control variables set to A = 540.25, B = 93, C = 18, D = 2, G = 50 and H = 50.5. Further, to achieve the maximization of MRR, the control variables were predicted and set as A = 540.25, B = 93, C = 18, E = 10.5, F = 1.5 and G = 50.5, as shown in Figure 2.14.

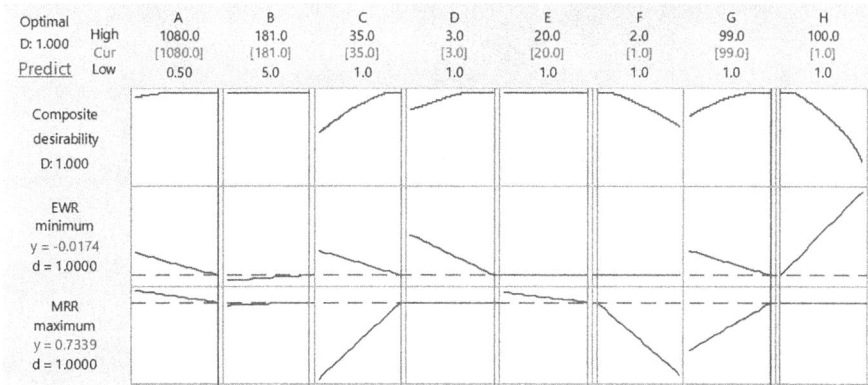

FIGURE 2.12 Response optimizer for EWR and MRR using regression analysis.

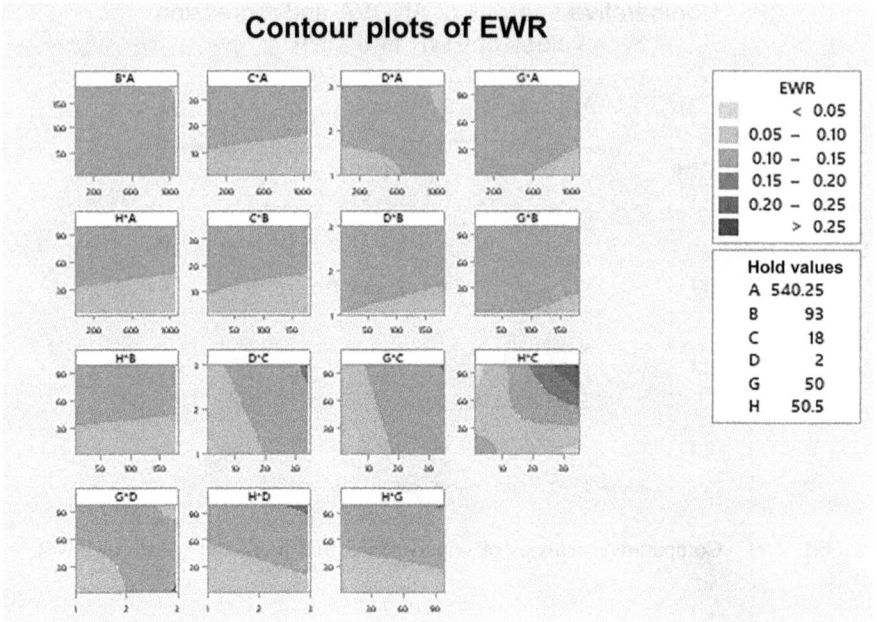

FIGURE 2.13   Contour plot for EWR obtained from regression analysis.

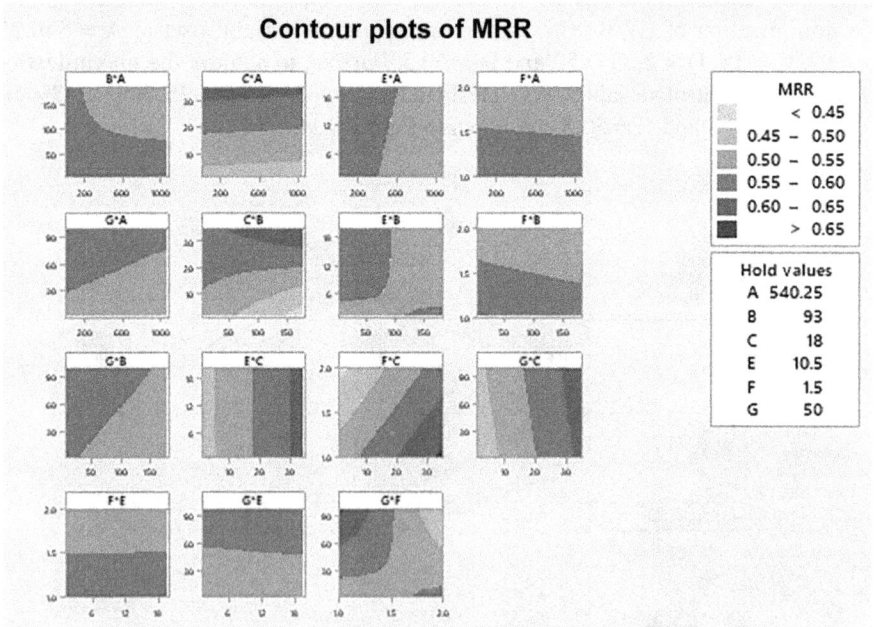

FIGURE 2.14   Contour plot for MRR obtained from regression analysis.

TABLE 2.10

Performance Improvement in EWR and MRR in the Novel Curved EDM Process

	Experimental Values	Performance Improvement Optimization			
		ANOVA Technique		Regression Analysis	
Levels	-	A1B1C1D2-E1F1G2H1	95% CI	A1B1C1D2-E1F1G2H1	95% CI
EWR (mm³/min)	0.11	−0.0174	(−0.2584, 0.2237)	−0.0174	(−0.2584, 0.2237)
MRR (mm³/min)	0.55	0.763	(0.4154, 1.1105)	0.7339	(0.4625, 1.0053

The nature-based advanced Jaya algorithm was implemented for obtaining the optimum values. The multi-objective optimization predicts the optimum control variables required for machining curved channels using the novel curved EDM process. The machining responses also played an important role in minimizing the EWR and maximizing the MRR. The regression equations (Eqs. 2.5 and2.6) were used as the objective functions for Jaya algorithm. The population size in Jaya algorithm was fixed at the beginning and is maintained constant in every generation throughout the simulation run. The set of solutions obtained using Jaya algorithm in a single simulation run for EWR and MRR are shown in Table 2.10. From the optimum result of machining variables in EWR, it was found that the value of pulse ON time (880.385µs) and bi-pulse current (1.758A), sparking time (14.939µs) was towards the maximum. Similarly, the average contribution of duty cycle (97.310µs) and lifting time (1.539µs) improves the EWR. The minimum value of sparking current (1.758A) and sensitivity (25.743 mm/min) reduces the EWR. The contribution of gap voltage (40.005 V) approaches towards the average value that improves the EWR. It was found that the average value of pulse ON time (580.886 µs), lifting time (1.244 µs) and sensitivity (41.447 mm/min) significantly contributes to MRR in the curved EDM process. A higher value of duty cycle (157.701 µs), sparking current (32.514A), bi-pulse current (2.789A), sparking time (15.824) and gap voltage (91.142 V) increases the MRR.

2.5   CONCLUSIONS

This chapter successfully established a mechanism for machining curved channels using the novel curved EDM process. The design and development with the standard operating procedure was discussed. Further, the experiments were formulated using the Taguchi orthogonal array (L12). The curved EDM process utilized a copper curved electrode and OHNS workpiece material. In total, eight control variables were identified: pulse ON time, duty cycle, sparking current, bi-pulse current, sparking time, lifting time, gap voltage and sensitivity, along with two machining responses, i.e. EWR and MRR. The main objective was to minimize the EWR and maximize the MRR in the novel curved EDM process. The major findings of this research are summarized as follows:

- The curved channel was generated and primarily finds its applications in injection mould for reducing the heat generation in plastic products.
- The major control variables affecting the EWR and MRR in the novel curved EDM process are sparking current (EWR, $p = 0.06$, and MRR, $p = 0.043$), sensitivity (EWR, $p = 0.064$) and the interaction of sparking current and sensitivity (EWR, $p = 0.059$) based on the p-value.
- In ANOVA and multiple regression analysis, there was a negligible contribution of sparking time and lifting time to EWR and of bi-pulse current and sensitivity to MRR, and hence, these control variables were eliminated. The remaining selected machining control variables with defined levels majorly contributed to the machining responses.
- The implementation of ANOVA tool and multiple regression analysis in the curved machining process was successfully validated using a response optimizer, resulting in the machining responses and control variables combinations of A1B1C1D2E1F1G2H1 for EWR and A1B1C1D2E1F1G2H1 for MRR.
- Jaya algorithm was used for identifying the machining variable with optimum machining responses (minimum EWR = 0.4445 mm³/min; maximum MRR = 0.779 mm³/min).
- The experimental values of EWR were minimized to 24.71%, and those of MRR maximized to 70.60%, respectively, with the optimum control variable combination for the performance improvement within the range of 95% confidence interval.

REFERENCES

1. N.M. Abbas, D.G. Solomon, and Md.F. Bahari, A review on current research trends in electrical discharge machining (EDM). *International Journal of Machine Tools & Manufacture*, 47, 1214–1228, 2007.
2. N. Malhotra, S. Rani, and H. Singh, Improvements in performance of EDM-A review. *IEEE Southeast Con 2008*, Huntsville, AL, pp. 599–603, 2008.
3. N.K. Singh, P.M. Pandey, K.K. Singh, and M.K. Sharma, Steps towards green manufacturing through EDM process: A review. *Cogent Engineering*, 3, 1, 2016.
4. A.Kr. Singh, R. Mahajan, A. Tiwari, D. Kumar, and R.K. Ghadai, Effect of dielectric on electrical discharge machining: A review. *IOP Conference Series: Materials Science and Engineering*, 377, 012184, 2018.
5. T. Ishida, and Y. Takeuchi, L-shaped curved hole creation by means of electrical discharge machining and an electrode curved motion generator. *The International Journal of Advanced Manufacturing Technology*, 19, 260–265 2002.
6. T. Ishida, and Y. Takeuchi, Creation of U-shaped and skewed holes by means of electrical discharge machining using an improved electrode curved motion generator. *International Journal of Automation Technology*, 2(6), 439–446, 2008.
7. M. Kita, T. Ishida, K. Teramoto, and Y. Takeuchi, Size reduction and performance improvement of automatic discharge gap controller for curved hole electrical discharge machining. In: K. Shirase, and S. Aoyagi (eds.) *Service Robotics and Mechatronics*. Springer, London, 2010.

8. P.K. Shrivastava, and A.K. Dubey, Electrical discharge machining–based hybrid machining processes: A review, *Proceedings of the Institution of Mechanical Engineers, Part B: Journal of Engineering Manufacture*, 228(6) 799–825, 2013.
9. A. Yamaguchi, A. Okada, T. Miyake, and T. Ikeshima, Development of curved hole drilling method by EDM with suspended ball electrode, *Journal of the Japan Society for Precision Engineering*, 81(11), 1039–1044, 2015.
10. D. Soaita, Electrode-tool for making holes on helical trajectories by EDM, *Procedia Technology*, 22, 74–77, 2016.
11. https://www.hsk.co.jp/english/introduction/electrical_discharge/magari_ana_process. html. Hoden Semitsu Kako Kenkyusho Co., Ltd, Japan (Last accessed on 2020 Jan 20).
12. D.B. Meshram, and Y.M. Puri, Review of research work in die sinking EDM for machining curved hole, *Journal of the Brazilian Society of Mechanical Sciences and Engineering*, 39, 2593–2605, 2017.
13. C. Lin, J. Lin, and T. Ko, Optimisation of the EDM process based on the orthogonal array with fuzzy logic and grey relational analysis method. *The International Journal of Advanced Manufacturing Technology*, 19, 271–277, 2002.
14. P.M. George, B.K. Raghunath, L.M. Manocha, and A.M. Warrier, EDM machining of carbon–carbon composite—A Taguchi approach. *Journal of Materials Processing Technology*, 145(1), 66–71, 2004.
15. T. Yuvaraj, and P. Suresh, Analysis of EDM process parameters on inconel 718 using the grey-taguchi and topsis method. *Strojniški vestnik - Journal of Mechanical Engineering*, 65(10), 557–564, 2019.
16. P.H. Nguyen, T.L. Banh, K.A. Mashood, D.Q. Tran, V.D. Pham, T. Muthuramalingam, V. D. Nguyen, and D.T. Nguyen, Application of TGRA-based optimisation for machinability of high-chromium tool steel in the EDM process. *Arabian Journal for Science and Engineering*, 2020. doi:10.1007/s13369-020-04456-z.
17. C.M. Judd, G.H. McClelland, and C.S. Ryan, *Data Analysis: A Model Comparison Approach to Regression, ANOVA, and Beyond*, 3rd Edition, Routledge, Createspace Independent Publishing Platform, Abingdon, 2017.
18. U. Aich, and S. Banerjee, Modeling of EDM responses by support vector machine regression with parameters selected by particle swarm optimization. *Applied Mathematical Modelling*, 38(11–12), 2800–2818, 2014.
19. S. Debnath, R.N. Rai, and G.R.K. Sastry, A study of multiple regression analysis on die sinking edm machining of ex-situ developed Al-4.5cu-SiC composite. *Materials Today: Proceedings*, 5(2), 5195–5201, 2018.
20. D.B. Meshram, and Y.M. Puri, Effective parametric analysis of machining curvature channel using semicircular curved copper electrode and OHNS steel workpiece through a novel curved EDM process. *Engineering Research Express*, 1, 015014, 2019.
21. R. Venkata Rao, Jaya: A simple and new optimization algorithm for solving constrained and unconstrained optimization problems. *International Journal of Industrial Engineering Computations*, 7, 19–34, 2016.
22. N. Agarwal, M.K. Pradhan, and N. Shrivastava, A new multi-response Jaya Algorithm for optimisation of EDM process parameters, *Materials Today: Proceedings*, 5(11), 23759–23768, 2018.
23. A. Khachane, and V. Jatti, Multi-objective optimization of EDM process parameters using jaya algorithm combined with grey relational analysis. In: Venkata Rao, R., and Taler J. (eds.) *Advanced Engineering Optimization Through Intelligent Techniques. Advances in Intelligent Systems and Computing*, 949. Springer, Singapore, 2020.
24. R.V. Rao, D.P. Rai, J. Ramkumar, and J. Balic, A new multiobjective Jaya algorithm for optimization of modern machining processes. *Advances in Production Engineering and Management*, 11(4), 271–286, 2016.

25. A. Majumder, P.K. Das, A. Majumder, and M. Debnath, An approach to optimize the EDM process parameters using desirability-based multi-objective PSO. *Production & Manufacturing Research*, 2(1), 228–240, 2014.
26. https://www.minitab.com/en-us/products/minitab/free-trial/ Minitab (Version 17) [Software] (Last accessed on 2020 Apr 24), 2013.
27. Puthumana, G., Micro-EDM process modeling and machining approaches for minimum tool electrode wear for fabrication of biocompatible micro-components, *Journal of Machine Engineering*, 17(3), 97–111, 2017.
28. Rao, R., Jaya: A simple and new optimization algorithm for solving constrained and unconstrained optimization problems, *International Journal of Industrial Engineering Computations*, 7(1), 19–34, 2016.
29. M. Das, R. Rudrapati, N. Ghosh, and L. Rathod, Input Parameters optimization in EDM process using RSM and JAYA Algorithm. *International Journal of Current Engineering and Technology*, 6, 109–112, 2016.

3 Artificial Neural Networks (ANNs) for Prediction and Optimization in Friction Stir Welding Process
An Overview and Future Trends

*Mukuna Patrick Mubiayi and
Veeredhi Vasudeva Rao*
University of South Africa

CONTENTS

3.1 FRICTION STIR WELDING (FSW) PROCESS

To fabricate some parts and components, welding is required. Welding is a process that enables joining of parts or components to produce various products of different geometries and complex shapes that sometimes are not easy to fabricate using other manufacturing techniques. There are different types of welding such as gas welding, resistance welding, and solid-state welding. In this chapter, the friction stir welding (FSW) technique is investigated to establish the current state of the usage of artificial neural networks (ANNs) to predict the properties of the fabricated welds or to optimize the process parameters of this type of joining technique. Solid-state welding is a technique where the joint is fabricated below the melting temperature of the parent materials and that takes place without the use of filler metal. There are many types of solid-state welding, which include friction stir welding, forge welding, explosive welding, ultrasonic welding, and friction welding.

As mentioned above, the FSW technique is one of the solid-state welding methods and was invented and patented in 1991 in the UK by The Welding Institute (TWI) (Thomas et al., 1991). The technique was originally used for butt and lap weld configurations of many types of materials, including ferrous and non-ferrous metals and plastic materials. The technique has evolved over the years and has been used to join a variety of materials. The joining technique uses a shaped rotating tool that is plunged between two abutted materials in case of a butt configuration. The tool in motion (rotation) and the relative motion of the materials to be welded generate frictional heat, which leads to the creation of a plasticized region around the plunged part of the rotating tool (Thomas et al., 1991). The joining process is usually described in four steps: plunging (the tool in rotation is plunged into the workpieces), dwelling (the tool remains at its position for a set time), welding (the tool moves forward along the weld line), and retraction (when the weld is complete, the tool is pulled out of the material). Welding problems including porosity, solidification cracking, and liquation cracking are eliminated when friction stir welding is used, and these are the advantages of FSW compared to fusion welding techniques. This is because of the solid-state nature of friction stir welding (Lienert et al., 2003). Figure 3.1 portrays a drawing of the friction stir welding technique. Mishra and Mohaney (2007) indicated that FSW has many benefits such as the usage of very less energy when compared to the energy needed for a laser weld (only 2.5%). Furthermore, FSW has many areas of applications, including aerospace (Hartley et al., 2002), automotive (Kallee, 2000), and robotics (Smith et al., 2003). This indicates that the technique is utilized in many industries and provides more advantages compared to other joining techniques.

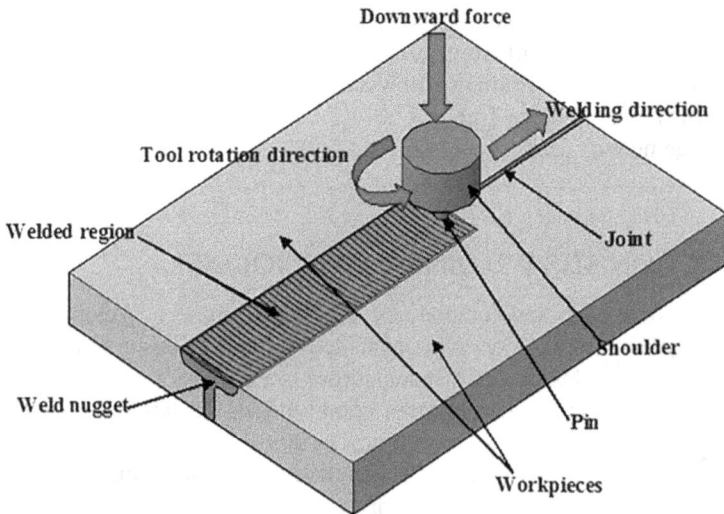

FIGURE 3.1   Friction stir welding (FSW) joining process.

Furthermore, the friction stir welding technique has many environmental benefits. Some of the environmental benefits are as follows (Mishra and Mohaney, 2007):

- Shielding gas is not necessary.
- Minimal surface cleaning is required.
- The grinding wastes are eliminated.
- The solvents required for degreasing are eliminated.
- The process saves consumable materials (i.e. rugs, wire).
- There are no harmful emissions.

3.1.1 FSW Process Parameters

The FSW technique uses a variety of process parameters, including the tool design, welding parameters (i.e. rotational speed, traverse speed), joint configurations (i.e. butt, lap), and tool forces (acting during the welding process, i.e. axial force, traverse force). The aforementioned process parameters have been shown to have significant effects on the microstructure evolution and mechanical properties, such as tensile strength and microhardness (Xue et al., 2011; Bisadi et al., 2013; Muthu and Jayabalan, 2015; Mehta and Badheka, 2015) of the fabricated welds. Figure 3.2 shows the representation of a FSW tool with basic geometry. Most of the tools used in FSW follow the same basic trends in relation to their shapes and geometries and usually comprise of a shoulder, a pin also called a probe, and any other external features. Furthermore, Mishra and Mohaney (2007) stated that the tool has three principal functions, namely the heating of the plates, the movement of the material to fabricate the weld, and the restraint of the hot metal under the tool shoulder.

In friction stir welding, there is a formation of different microstructural zones, namely heat-affected zone (HAZ), thermomechanically affected zone (TMAZ), and stir zone (SZ), and the unaffected material is called base material (Mishra and Mohaney, 2007).

The base material is the part far from the welded zone that has not been deformed and whose microstructure and mechanical properties are not affected by the heat from the welding process. The heat-affected zone (HAZ) is the area closer to the centre of the

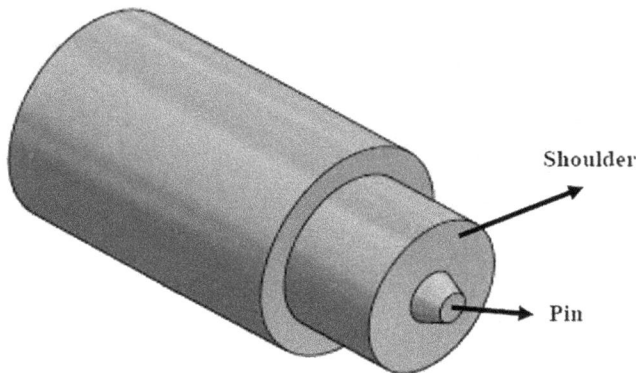

FIGURE 3.2 Geometry of typical FSW tool.

joint where the material is affected by a thermal cycle, resulting in the modification of the microstructure and/or the mechanical properties. Furthermore, in the HAZ, there is no occurrence of plastic deformation (Mishra and Mohaney, 2007). On the other hand, the thermomechanically affected zone (TMAZ) is situated in the area where the tool has produced material plastic deformation and the generated heat from the welding has produced some influence on the material. In the stir zone (weld nugget), full recrystallization takes place in the area previously occupied by the tool pin (Mishra and Mohaney, 2007). Friction stir welding technique has been used to join a large variety of materials in similar and dissimilar joint configurations (aluminium, copper, magnesium, titanium, and steel). Furthermore, the prediction of joint properties and the optimization of process parameters will broaden the application of this joining technique.

3.2 ARTIFICIAL NEURAL NETWORKS (ANNs)

Artificial neural networks (ANNs) are computational network tools used in an attempt to simulate the decision process in networks of neurons (nerve cell) of the biological central nervous system of humans or animals (Daniel, 2013). In simple words, ANNs can be defined as an imitation of the human brain showing its ability to learn new things and adapt to a new environment and the changes in the same. Figure 3.3 portrays a biological neural cell called neuron, which is a component of a biological neural network.

The signal between neurons is received from the dendrites, and when the signal surpasses a certain threshold, the neuron activates its own signal and that signal is passed to the next neuron. This manoeuvre is done via the axon, and a large number of neurons work concurrently (Gershenson, 2003).

In 1943, McCulloch and Pitts were the first to present the basic principles of the artificial neural networks (ANNs) in the five following assumptions (Daniel, 2013):

1. The activity of a neuron is uncompromising.
2. For the neuron to be excited, it is necessary that a certain fixed number of synapses larger than one be excited within a given interval of neural addition.

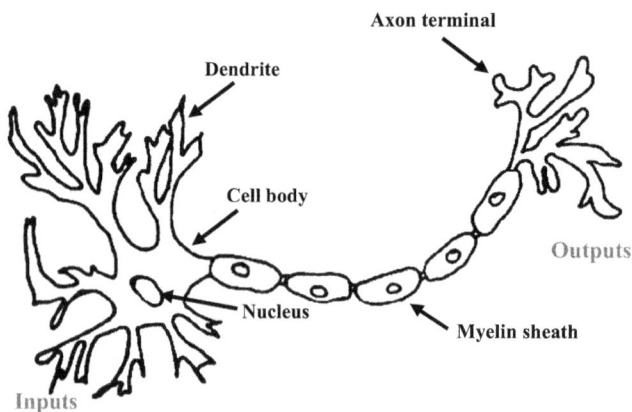

FIGURE 3.3 Biological neural cell (neuron).

3. The synaptic delay is the only significant delay in the neural system.
4. The excitation of the neuron is prevented by the activity of any inhibitory synapse at that time.
5. Over time, the structure of the interconnection network does not change.

In 1958, the psychologist Frank Rosenblatt created the first artificial neural network (ANN) and called it 'perceptron' (Dewan et al., 2016). Furthermore, artificial neural networks are often used to model complex nonlinear functions in numerous applications (Dewan et al., 2016). An artificial neural network architecture consists of three layers, namely the input layer, the hidden layer, and the output layer. The chosen input factors are put in the input layer, the information from the input is processed in the hidden layer(s), and then the result is obtained from the output layer (Dewan et al., 2016). Figure 3.4 presents an ANN architecture consisting of four layers: an input layer, two hidden layers, and one output layer. The main architecture of artificial neural networks (ANNs), when taking into consideration the disposition of the neuron, how the ANNs are interconnected, and how ANNs layers are composed, can be divided in the following way (da Silva et al., 2017):

- Single-layer feedforward networks.
- Multi-layer feedforward networks.
- Recurrent networks.
- Mesh networks.

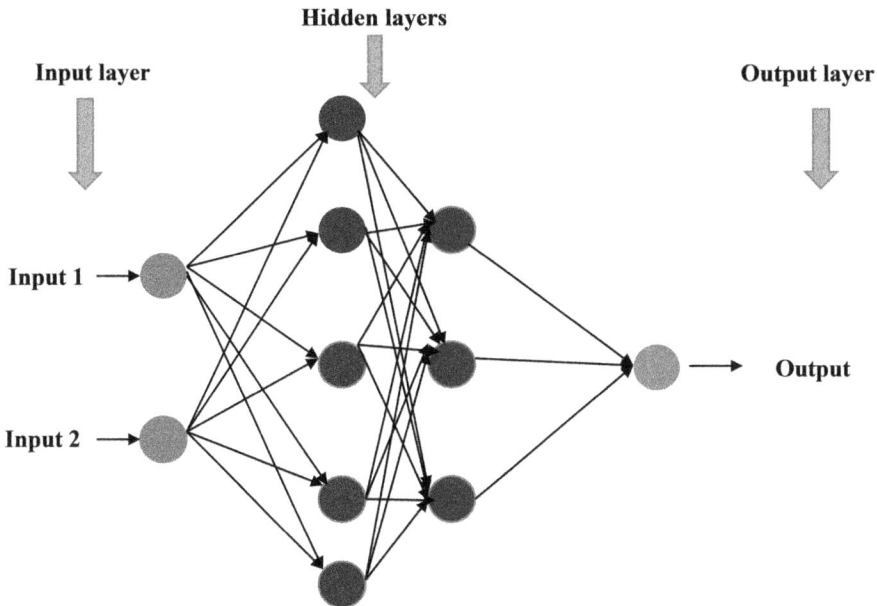

FIGURE 3.4   An artificial neural network (ANN) diagram and its layers.

More details on the main architectures of the artificial neural networks (ANNs) can be found in da Silva et al. (2017). In ANNs, many algorithms are used for the training of neurons, such as backpropagation, gradient descent with momentum, and Levenberg–Marquardt algorithms.

It was observed that the backpropagation algorithm is the most used method in training the ANN for the prediction of the properties of the friction stir welds. In this chapter, a background on ANNs is provided without providing details on the mathematical aspect of ANNs. Only the fundamental abilities of artificial neural networks are presented, and more details on ANNs can be found in Daniel (2013).

The ANN shown in Figure 3.4 comprises of two input variables, two hidden layers with 4 and 3 nodes, and one output.

3.2.1 Applications of ANNs

Artificial neural networks have effectively been utilized to solve a variety of optimization problems in real-world applications, and their popularity has considerably increased over the years. ANNs are considered as effective tools for resolving optimization problems due to their good properties such as flexibility, high efficiency, and simplicity. Table 3.1 depicts some initial works related to ANN models used for predictions and optimizations, with the year of invention and the creators. Over the years, ANNs have successfully been used in many disciplines, including engineering (Mashhadban et al., 2016; Al Shamisi et al., 2011), economics (Kim et al., 2004; Falat and Pancikova, 2015; Badea, 2014), and medicine

TABLE 3.1
ANN Models Utilized for Prediction and Optimization

ANN Model	Inventor	Year	Application	References
Perceptron networks	Rosenblatt	1958	Prediction	Rosenblatt (1958) as cited in Bermejo et al. (2019)
Adaline/Madaline	Bernard Widrow	1960	Prediction	Widrow (1960) as cited in Bermejo et al. (2019)
Hopfield networks	Hopfield	1982	Optimization	Chen and Amarias (2001) as cited in Bermejo et al. (2019)
Backpropagation	Rumelhart and Parker	1985	Prediction	Rumelhart et al. (1985) as cited in Bermejo et al. (2019)
Digital neural network architecture (DNNA)	Neural Semiconductor Inc.	1990	Prediction	Tomlinson et al. (1990) as cited in Bermejo et al. (2019)

Furthermore, it is expected that the usage of ANNs modelling tool would mitigate the presence of welding defects, such as voids, pores and cracks. Furthermore, the optimization of the FSW process parameters could extend its industrial applications and have a positive impact on the environment.

(Shahid et al., 2019; Patel and Goyal, 2007), and substantial progress has been made over the past few years.

It can be said that, in the engineering field, ANNs have been used for optimizations and predictions (Mashhadban et al., 2016; Al Shamisi et al., 2011; Tansel et al., 2010; Dewan et al., 2016; Dehabadi et al., 2016). Based on the broad usage of ANNs in engineering, it could be said that many other engineering processes can use this optimization tool to reduce the number of trial-and-error experiments carried out in various industries. This could lead to the reduction in time taken for various processes to be carried out. It was observed that for the predictions and optimizations in the FSW technique, the backpropagation (BP) algorithm is the most used. Rumelhart, Hinton, and Williams in 1986 proposed the backpropagation algorithm for setting weights and, therefore, for the training of multi-layer perceptrons (Daniel, 2013).

3.3   ANN UTILIZATION IN FRICTION STIR WELDING

The usage of artificial neural networks (ANNs) in engineering fields including manufacturing has yielded good results in process monitoring, prediction, and optimization. In friction stir welding (FSW), researchers have conducted studies for properties prediction and process optimization using ANNs. In FSW, artificial neural networks have been used for properties prediction (i.e. microhardness, tensile strength) and optimization (i.e. rotation speed, welding speed) (Okuyucu et al., 2007; Tansel et al., 2010; Dehabadi et al., 2016; Palanivel et al., 2016; Babu et al., 2018). Table 3.2 gives details on the type of algorithm, the utilization, the input and output of the ANN structure used in studies on friction stir welding.

Over the years, artificial neural networks (ANNs) have successfully been used to solve prediction and optimization problems in various fields, including friction stir welding (FSW). Therefore, it is noteworthy to summarize some studies conducted in the field of FSW using artificial neural networks. In their research, Buffa et al. (2012) used a neural network to predict the microhardness and microstructural properties of 3-mm-thick Ti–6Al–4V friction stir welds. The trained networks were linked to a finite element model to obtain a unique tool capable of providing together with the local values of the main field variables. The obtained information from the tool was then used to set the process parameters used to conduct the experiments. The results showed a considerable agreement between the experiment and the prediction for the microhardness, whereas for the microstructure, an excellent prediction was achieved as depicted in Figure 3.5a and b.

AA8014 aluminium alloy was friction-stir-welded by Ghetiya and Patel (2014). An artificial neural network using a backpropagation algorithm was used to predict the tensile strength properties. The experimental data were in good agreement with the predicted data, and the error in the measured and predicted values was very low (less than 3%) for the 4–8–1 artificial neural network architecture (4 neurons in the input layer, 8 neurons in the hidden layer, and 1 neuron in the output layer). Hence, Ghetiya and Patel (2014) indicated that the developed ANN can be utilized as a different way for the prediction of tensile strength using specific process parameters.

TABLE 3.2

ANN Algorithms, Utilization, Input, and Output of Selected Studies in Friction Stir Welding

Algorithms	Utilization	Input	Output	References
Backpropagation	Prediction	Rotational speed, welding speed, pin length, axial force, tool shoulder, pin radius	Peak temperature, total torque, traverse force, bending stress, maximum shear stress	Manvatkar et al. (2012)
Backpropagation	Prediction	Weld speed, tool rotation speed	Hardness of the metal, hardness of HAZ, elongation, yield strength, tensile strength	Okuyucu et al. (2007)
Backpropagation	Prediction	Tool tilt angle, thread, distance from welding centreline	Microhardness	Dehabadi et al. (2016)
Backpropagation	Prediction	Tool shoulder diameter, tool rotational speed, welding speed, axial force	Tensile strength	Ghetiya and Patel (2014)
Backpropagation	Prediction	Rotation speed, welding speed	Ultimate tensile strength and hardness	Shojaeefard et al. (2013)
Backpropagation	Prediction	Plastic strain, strain rate, temperature, Zener–Hollomon parameter	Average grain size	Fratini et al. (2009)
Backpropagation	Prediction	Traverse speed, rotational speed	Peak temperature, HAZ area, welding force	Shojaeefard et al. (2014)
Gradient descent with momentum and Levenberg–Marquardt	Prediction	Weld speed, tool rotation speed	Tensile strength, yield strength, elongation	Yousif et al. (2008)

(*Continued*)

TABLE 3.2 (*Continued*)
ANN Algorithms, Utilization, Input, and Output of Selected Studies in Friction Stir Welding

Algorithms	Utilization	Input	Output	References
Backpropagation	Prediction	Tool pin profile, rotational speed, welding speed, axial force	Ultimate tensile strength	Palanivel et al. (2016)
Backpropagation	Prediction	Plastic strain, strain rate, temperature	Microhardness, microstructure	Buffa et al. (2012)
Backpropagation	Prediction	Spindle speed, plunge force, welding speed	Ultimate tensile strength	Dewan et al. (2016)
Backpropagation	Prediction and optimization	Rotation speed, welding speed (for prediction). Tensile strength, yield strength, elongation, hardness, hardness of HAZ (for optimization)	Tensile strength, yield strength, elongation, hardness, hardness of HAZ (prediction). Rotation speed, welding speed (for optimization)	Tansel et al. (2010)
Batch backpropagation, quick propagation, incremental backpropagation	Prediction and optimization	Rotation speed, welding speed, tool tilt angle (for prediction)	Ultimate tensile strength, microhardness, corrosion resistance (for prediction). Rotation speed, welding speed, tool tilt angle (for optimization)	Babu et al. (2018)

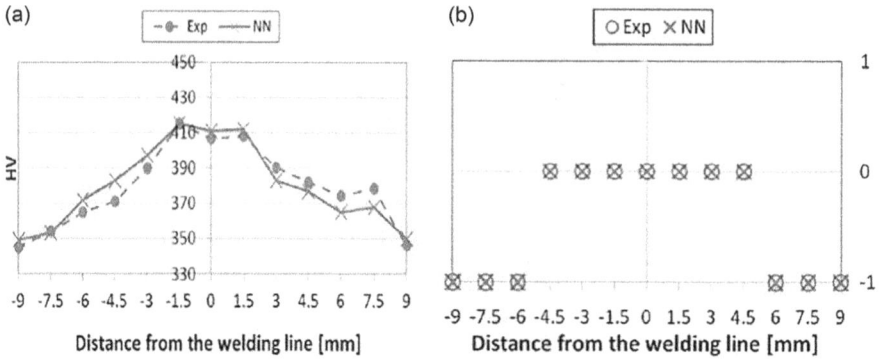

FIGURE 3.5 Experimental (Exp) and predicted (NN) (a) microhardness values and (b) microstructure at 1.5 mm distance from the bottom of the weld (Buffa et al., 2012). (Comment: Reprinted with permission from Elsevier.)

Okuyucu et al. (2007) investigated the possibility of using ANN for the prediction of mechanical properties (tensile strength, hardness of the metal and the heat-affected zone (HAZ), elongation, and yield strength). They found small error values for the outputs, and the R^2 were all bigger than 0.99 except for the elongation. They further found that the correlations between the measured and the predicted values were better, except for the elongation and yield strength. The microhardness profiles of AA6061 friction stir welds were obtained in different distances from the weld line. The mean absolute percentage error (MAPE) of the artificial neural networks was less than 4.83%, indicating that the difference between the experimental and the predicted microhardness values is acceptable. They concluded that the usage of mathematical modelling methods such as ANN can reduce time, material, and cost and results in optimized designs (Dehabadi et al., 2016).

Babu et al. (2018) combined artificial neural network with genetic algorithm to model the connection between the rotation speed, traverse speed, and tool tilt angle (input) and the microhardness, tensile strength, and corrosion resistance (output). Furthermore, ANN-GA was used to optimize the process parameters. They found that the method reduces the number of trial runs and the associated costs. Therefore, it can be said that the usage of ANN and GA (genetic algorithm) is an excellent tool to use for cost reduction in friction stir welding process.

Tansel et al. (2010) developed a genetically optimized neural network system abbreviated as GONNS to model the weld properties and optimize the welding process parameters. The GONNS is the combination of ANN and genetic algorithm (GA), and it is used to model the link between the inputs and the outputs of the considered structure. They found that GONNS is a practicable option for modelling and optimizing the friction stir welding process. They further recommended the usage of the GONNS because it worked as expected in the application. Figure 3.6a and b shows a model based on five artificial neural networks and a genetically optimized neural network structure.

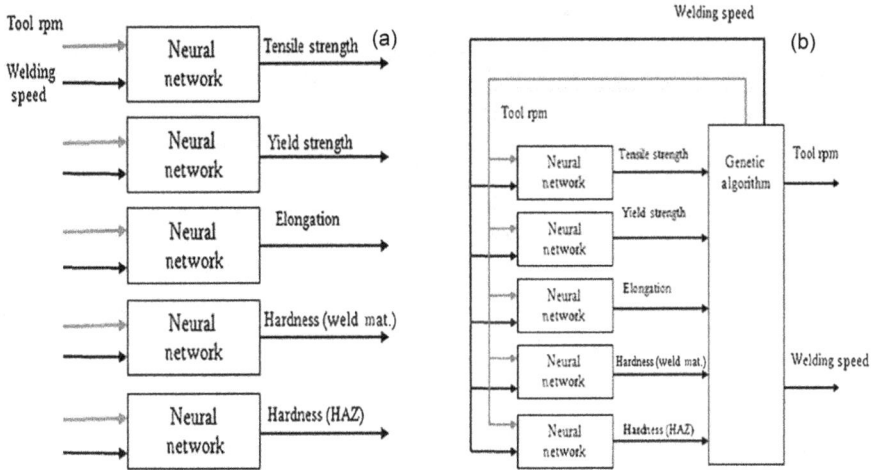

FIGURE 3.6 (a) A five-ANN structure; (b) a GONNS (Tansel et al., 2010). (Comment: Reprinted with permission from Springer.)

On the other hand, Shojaeefard et al. (2014) used a backpropagation algorithm to predict the peak temperature and the heat-affected zone (HAZ) width and the welding force for a series of rotation and welding speeds. They stated that the ANN performance was excellent and can successfully be used to predict the HAZ width, peak temperature, and welding force for friction stir welds fabricated using a range of rotation and welding speeds. Other studies also used ANN for the prediction (output) of various properties ranging from tensile strength (Palanivel et al., 2016; Dewan et al., 2016), total torque, peak temperature, traverse force, bending stress, maximum shear stress (Manvatkar et al., 2012), and average grain size (Fratini et al., 2009). And these used various inputs, including traverse speed, rotation speed, pin radius, pin length, and axial force. The tool tilt angle, plastic strain, temperature, and strain rate have also been used as input in some studies.

It was observed that most of the studies carried out by researchers used ANNs to predict the resulting properties of the fabricated friction stir welds. Furthermore, very few used a combination of the artificial neural network and genetic algorithm to optimize the welding process parameters. It was further observed that most of the case studies in the open literature on friction stir welding using ANNs are on the welding of aluminium alloys.

3.4 CONCLUSION AND FUTURE TRENDS

There are many prediction and optimization techniques which are used to generate a correlation between the outputs and the process inputs through the developing mathematical models. Furthermore, in the open literature, many studies are linked to the applications of artificial intelligence (AI), predominantly artificial neural networks (ANNs), in many engineering fields, including manufacturing and construction. Techniques such as response methodology (RSM) and artificial neural networks (ANNs) are used to understand the relationship between the process input parameters and the resulting output. It was observed that in the open literature, only a limited number of base materials on FSW are investigated using the artificial neural networks as a modelling tool.

Over the years, many researchers in numerous sectors, including engineering, economics, and medicine, have shown interest in using the artificial neural networks. They successfully used ANNs to solve a range of problems in their areas of research. Most of the studies summarized in this chapter showed that ANNs are a suitable and powerful tool to use in the prediction of welding properties. It can then be concluded that ANNs are a very useful model and can be used for the prediction of the properties of friction stir welds, which would lead to solving problems occurring in this joining technique and could also lead to the broader usage of the technique. Furthermore, it was also observed that many studies on friction stir welding where ANN modelling tool has been used are on aluminium. The usage of artificial neural networks should be extended to other joint configurations, especially to dissimilar joint configurations. However, to better utilize artificial neural networks for solving different problems, it is important to understand the potential and the limitations of the ANNs.

ACKNOWLEDGEMENTS

The authors would like to acknowledge the support from the University of South Africa (South Africa). The permissions from various publishers to use figures are also acknowledged.

REFERENCES

Al Shamisi, M.H., Assi, A.H. and Hejase, H.A., 2011. *Using MATLAB to Develop Artificial Neural Network Models for Predicting Global Solar Radiation in Al Ain City-UAE.* INTECH Open Access Publisher, Rijeka.

Badea, L.M., 2014. Predicting consumer behavior with artificial neural networks. *Procedia Economics and Finance*, 15, pp. 238–246.

Babu, K.K., Panneerselvam, K., Sathiya, P., Haq, A.N., Sundarrajan, S., Mastanaiah, P. and Murthy, C.S., 2018. Parameter optimization of friction stir welding of cryorolled AA2219 alloy using artificial neural network modeling with genetic algorithm. *The International Journal of Advanced Manufacturing Technology*, 94(9–12), pp. 3117–3129.

Bermejo, J.F., Fernández, J.F.G., Polo, F.O. and Márquez, A.C., 2019. A review of the use of artificial neural networks models for energy and reliability prediction: A study for the solar PV, hydraulic and wind energy sources. *Applied Science*, 9, p. 1844.

Bisadi, H., Tavakoli, A., Tour Sangsaraki, M. and Tour Sangsaraki, K., 2013. The influences of rotational and welding speeds on microstructures and mechanical properties of friction stir welded Al5083 and commercially pure copper sheets lap joints. *Materials and Design*, 43, pp. 80–88.

Buffa, G., Fratini, L. and Micari, F., 2012. Mechanical and microstructural properties prediction by artificial neural networks in FSW processes of dual phase titanium alloys. *Journal of Manufacturing Processes*, 14(3), pp. 289–296.

Daniel, G., 2013. *Principles of Artificial Neural Networks* (Vol. 7). World Scientific, Singapore.

da Silva, I.N., Spatti, D.H., Flauzino, R.A., Liboni, L.H.B. and dos Reis Alves, S.F., 2017. Artificial neural network architectures and training processes. In: *Artificial Neural Networks* pp. 21–28. Springer, Cham.

Dehabadi, V.M., Ghorbanpour, S. and Azimi, G., 2016. Application of artificial neural network to predict Vickers microhardness of AA6061 friction stir welded sheets. *Journal of Central South University*, 23(9), pp. 2146–2155.

Dewan, M.W., Huggett, D.J., Liao, T.W., Wahab, M.A. and Okeil, A.M., 2016. Prediction of tensile strength of friction stir weld joints with adaptive neuro-fuzzy inference system (ANFIS) and neural network. *Materials & Design*, 92, pp. 288–299.

Falat, L. and Pancikova, L., 2015. Quantitative modelling in economics with advanced artificial neural networks. *Procedia Economics and Finance*, 34, pp.194–201.

Fratini, L., Buffa, G. and Palmeri, D., 2009. Using a neural network for predicting the average grain size in friction stir welding processes. *Computers & Structures*, 87(17–18), pp. 1166–1174.

Gershenson, C., 2003. Artificial neural networks for beginners. arXiv preprint cs/0308031.

Ghetiya, N.D. and Patel, K.M., 2014. Prediction of tensile strength in friction stir welded aluminium alloy using artificial neural network. *Procedia Technology*, 14, pp. 274–281.

Hartley, P.J., Hartley, D.E. and McCool, A., 2002. FSW Implementation on the Space Shuttle's External Tank.

Kallee, S., 2000. Application of friction stir welding in the shipbuilding industry. Lightweight construction. The Royal Institution of Naval Architects, TWI, Cambridge, UK, p. 25.

Kim, T.Y., Oh, K.J., Sohn, I. and Hwang, C., 2004. Usefulness of artificial neural networks for early warning system of economic crisis. *Expert Systems with Applications*, 26(4), pp. 583–590.

Lienert, T.J., Stellwag, W.L., Grimmett Jr., B.B. and Warke, R.W., 2003. Friction stir welding studies on mild steel' Welding research, American Welding Society and the Welding Research Council.

Manvatkar, V.D., Arora, A., De, A. and DebRoy, T., 2012. Neural network models of peak temperature, torque, traverse force, bending stress and maximum shear stress during friction stir welding. *Science and Technology of Welding and Joining*, 17(6), pp. 460–466.

Mashhadban, H., Kutanaei, S.S. and Sayarinejad, M.A., 2016. Prediction and modeling of mechanical properties in fiber reinforced self-compacting concrete using particle swarm optimization algorithm and artificial neural network. *Construction and Building Materials*, 119, pp. 277–287.

Mehta, K.P. and Badheka, V.J., 2015. Influence of tool design and process parameters on dissimilar friction stir welding of copper to AA6061-T651 joints. *The International Journal of Advanced Manufacturing Technology*, 80(9–12), pp. 2073–2082.

Mishra, R.S. and Mahoney, M.W., 2007. Introduction. In: Mishra, R.S. and Mahoney, M.W. (eds.) *Friction Stir Welding and Processing*. ASM International, Materials Park, OH.

Muthu, M.F.X. and Jayabalan, V., 2015. Tool travel speed effects on the microstructure of friction stir welded aluminium - copper joints. *Journal of Materials Processing Technology*, 217, pp. 105–113.

Okuyucu, H., Kurt, A. and Arcaklioglu, E., 2007. Artificial neural network application to the friction stir welding of aluminum plates. *Materials & Design*, 28(1), pp. 78–84.

Palanivel, R., Laubscher, R.F., Dinaharan, I. and Murugan, N., 2016. Tensile strength prediction of dissimilar friction stir-welded AA6351–AA5083 using artificial neural network technique. *Journal of the Brazilian Society of Mechanical Sciences and Engineering*, 38(6), pp. 1647–1657.

Patel, J.L. and Goyal, R.K., 2007. Applications of artificial neural networks in medical science. *Current Clinical Pharmacology*, 2(3), pp. 217–226.

Shahid, N., Rappon, T. and Berta, W., 2019. Applications of artificial neural networks in health care organizational decision-making: A scoping review. PloS one, 14(2), e0212356.

Shojaeefard, M.H., Behnagh, R.A., Akbari, M., Givi, M.K.B. and Farhani, F., 2013. Modelling and Pareto optimization of mechanical properties of friction stir welded AA7075/AA5083 butt joints using neural network and particle swarm algorithm. *Materials & Design*, 44, pp. 190–198.

Shojaeefard, M.H., Akbari, M. and Asadi, P., 2014. Multi objective optimization of friction stir welding parameters using FEM and neural network. *International Journal of Precision Engineering and Manufacturing*, 15(11), pp. 2351–2356.

Smith, C.B., Hinrichs, J.F. and Crusan, W.A., 2003, Robotic friction stir welding: The state of the art. In *Proceedings of the Fourth International Symposium of Friction Stir Welding*, Utah, USA, 2003, pp. 14–16.

Tansel, I.N., Demetgul, M., Okuyucu, H. and Yapici, A., 2010. Optimizations of friction stir welding of aluminum alloy by using genetically optimized neural network. *The International Journal of Advanced Manufacturing Technology*, 48(1–4), pp. 95–101.

Thomas, W.M., Nicholas, E.D., Needham, J.C., Murch, M.G., Temple-Smith, P. and Dawes, C.J., 1991. Friction stir Butt Welding. International Patent No. PCT/GB92/02203, GB patent application No. 9125978.8.

Xue, P., Ni, D.R., Wang, D., Xiao, B.L. and Ma, Z.Y., 2011. Effect of friction stir welding parameters on the microstructure and mechanical properties of the dissimilar Al–Cu joints *Materials Science and Engineering A*, 528, pp. 4683–4689.

Yousif, Y.K., Daws, K.M. and Kazem, B.I., 2008. Prediction of friction stir welding characteristic using neural network. *Jordan Journal of Mechanical and Industrial Engineering*, 2(3), pp. 151–155.

4 Energy-Efficient Cluster Head Selection for Manufacturing Processes Using Modified Honeybee Mating Optimization in Wireless Sensor Networks

Pramod D Ganjewar
MIT Academy of Engineering, Alandi(D.),
Pune, Maharashtra, India

Barani S.
Sathyabama Institute of Science and Technology
(Deemed To be University), Chennai, Tamilnadu, India

Sanjeev J. Wagh
Government College of Engineering,
Karad, Maharashtra, India

CONTENTS

4.1   INTRODUCTION

Recent progress in the design of miniaturization has led to the growth of small battery-operated sensors capable of detecting environmental characteristics such as temperature and sound. Generally, sensors are fitted with the capacity for data processing and communication. The prospective use of WSNs, such as disaster management, fighting field recognition, border protection, and safety surveillance, has risen over the previous few years. Sensors are anticipated to be deployed remotely in big numbers in these applications and function independently in unattended settings. Because a WSN consists of nodes with a non-replaceable energy resource, the main concern is to prolong the network's life. There are a number of sensor nodes and a sink in a WSN. The research community has extensively pursued the grouping of sensor nodes into clusters to attain the goal of network scalability. A sensor node is chosen in each cluster, called the CH. The CH is accountable for the overall request, for obtaining sensed information from the other sensors in the same cluster, and for routing this information to the sink. The energy consumption at CHs is therefore greater than that of the other nodes in the network. The CH in the cluster is alternated between sensor nodes in order to reduce the energy consumption to elongate the network's life. The CH selection method will therefore influence the network's lifetime. In this chapter, we will boost the lifetime of the network by using honeybee optimization with energy distributed clustering (EDC-HBO) protocol. We will pick the high-energy cluster head with these algorithms, which will devote less energy to transmitting aggregated information to the base station. By minimizing the energy consumption, clustering is one of the best methods to extend the network's lifetime.

4.2   LITERATURE REVIEW

Jafarizadeh et al. [1] used naive Bayes classifiers to identify the strong cluster head to boost the network's life in a WSN. The parameters used for selecting the cluster head are the position, balance energy/power, distance from the sink, and class. The naive Bayes classifier with predefined simulation parameters was implemented, and the performance was assessed. The results achieved were better than those of LEACH algorithm. Zahedi et al. [2] used the concept of booking to reduce message transmission and energy dissipation. By using the booking system, the amount of message transmission can be decreased. The suggested reservation-based clustering strategy was quite good at reducing energy dissipation. It minimizes energy and extends network's life by adding stages of reservation at the time of network setup. Subsequently, energy is minimized by efficiently reducing the control messages in the network.

Vijayalakshmi and Anandan [3] proposed particle swarm optimization and tabu search technique. By the selection of ideal route, it boosts the network's life. The results showed that the quality of cluster formation was enhanced, the proportion of live nodes was also enhanced, and the packet loss rate and delay were lowered. The comparison of the suggested tabu search hybrid heuristic strategy and the LEACH-based particle swarm optimization showed that the LEACH multi-hop protocol is inefficient. Selvi et al. [4] used the method of honeybee optimization to improve the network's life, performance, node scalability, quality, and energy efficiency. The method used discovered an ideal way to reduce energy consumption at a low cost. Therefore, a biologically effective strategy was proposed based on bee colony-inspired building of energy clusters. But after execution, the speed of packet transmission was discovered to be greater than the other methods.

Sengottuvelan and Prasath [5] suggested an algorithm called artificial fish swarm optimization to optimize cluster head selection. The findings showed fast convergence, enhanced capacity of fault tolerance, and enhanced search for local optimization. This technique minimizes and prolongs the life of the network. Daflapurkar et al. [6] suggested a technique composed of three steps, namely the building of a hop tree in the choice of sensor nodes and cluster formation from end to end in the cluster head. This work used the shortest-path tree routing technique. The findings showed that the technique outperforms the current energy-efficient routing alternatives.

Jha and Eyong [7] proposed a technique for optimizing energy consumption. The findings showed that the battery life of the nodes was extended by the energy values acquired during communication. Murugan and Sarkar [8] proposed firefly cyclic grey wolf optimization. This focused on energy stabilization, minimizing the distance and the delay between two sensor nodes. Two algorithms firefly and grey wolf optimization were combined. The performance of the proposed technique was compared to genetic algorithm, group search optimization, artificial bee colony, fractional artificial bee colony, and cyclic randomization firefly for the selection of cluster head. The parameters used for the performance analysis of all the implemented algorithms were network life, energy efficiency, and dead node statistics. It was demonstrated that the suggested algorithm extended the network's life.

4.3 PROPOSED SYSTEM

In honeybee mating optimization (HBMO), initially, a sensor node is randomly chosen as the cluster head. On a regular basis, the sensor nodes send the information to the cluster head of the same cluster [9]. The cluster head is responsible for the aggregation of the information gathered from all the nodes in the cluster and sends that information to the base station. This operation is performed in many iterations. Every cluster will pick a fresh coordinator during every iteration. The selection rule is to randomly select a node from the same cluster: from the nodes that are never selected as a coordinator or from those that have minimal [10] time to be a coordinator. The technique does not consider the ability to manage the issue of lifetime.

We have therefore suggested an efficient cluster head selection method called modified honeybee algorithm, in which cluster heads are chosen based on their favorable parameters such as minimum node distance from the sink, minimum local distance, and maximum residual energy, instead of randomly choosing parameters.

The Euclidean distance is used to calculate the minimum range and to find out the cluster head that has the minimum distance from the base station or sink [11]. Then,

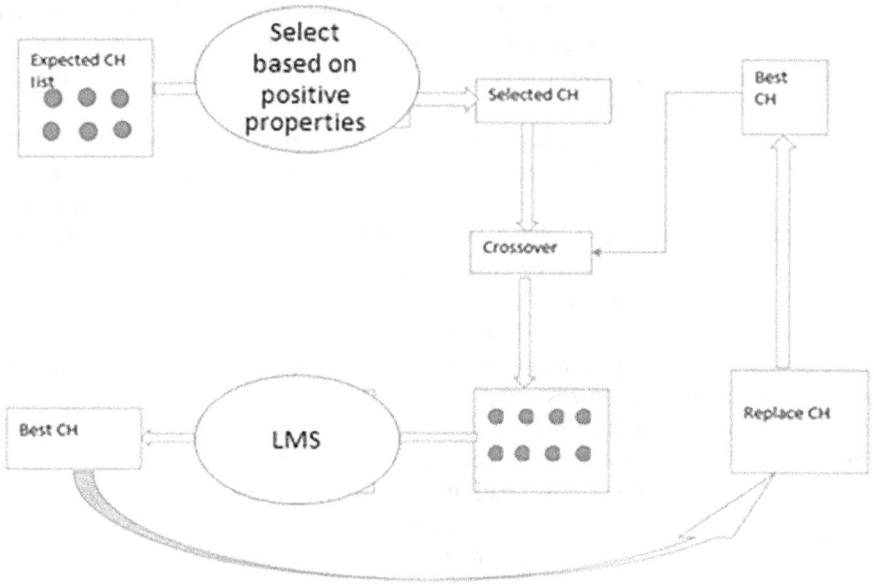

FIGURE 4.1   Modified honeybee mating optimization architecture.

the LMS classification algorithm is used instead of heuristic search, for better outcomes. The LMS classifier classifies the efficient cluster head from the cluster head
that was chosen earlier. The modified algorithm minimizes the energy utilization and
prolongs the network's life. The system architecture of HBMO is shown in Figure 4.1.

The major components of the modified honeybee mating optimization are
elaborated here.

4.3.1   Honeybee Optimization (HBO)

Nazir et al. [12] proposed an artificial bee colony (ABC) algorithm. The algorithm
uses the smart foraging behavior of honeybee swarms. This is a population-based,
simple, robust, and stochastic optimization technique. The artificial bee colony
algorithm has three groups of bees: employed bees, onlookers, and scouts. The
onlooker bees are responsible for selecting the good food sources from those found
by employed bees. The scout bee will randomly search for new sources. The position of food source is a possible solution for the optimization problem, and food
source's quantity of nectar corresponds to the solution's quality (fitness), as shown in
Figure 4.2.

4.3.2   Least Mean Squares (LMS) Classification

The LMS is a simple and popular algorithm. The update function in LMS is the multiplication of the step size, the current value of the error signal, and the input signal.
The update function does not depend on any other previous values.

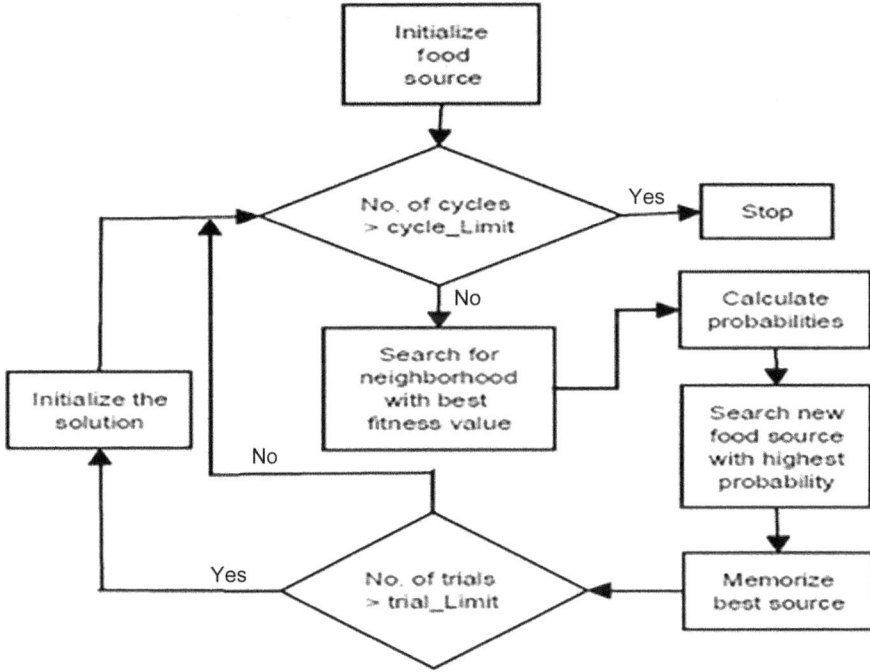

FIGURE 4.2 Honeybee classification.

4.3.3 MATHEMATICAL DESCRIPTION OF LMS AND ITS VARIANTS

The weight update in LMS and its variants is given by the following equation

$$w(n+1) = w(n) + \mu e(n)x(n) \qquad (4.1)$$

where
  $w(n + 1)$ – the value of new weight at time $(n + 1)$,
  $w(n)$ – the value of the weight at time n,
  μ – the step size and μ should be $0 \leq \mu \leq 2/\lambda_{\max}$

$$e(n) = d(n) - y(n) \qquad (4.2)$$

where

$$y(n) = \sum wn(k)x*(n-k)$$

4.4 IMPLEMENTATION

The details about the MHBMO algorithm, its pseudocode, and simulation parameters are given in this section.

4.4.1 Modified Honeybee Mating Optimization Algorithm

In the modified algorithm, instead of randomly selecting nodes for cluster head, we will be selecting nodes based on the following positive properties:

1. Minimum distance from the sink node.
2. Maximum remaining energy.
3. Minimum distance between two nodes.

Firstly, we will calculate the inter-node distance for cluster formation using Euclidian distance:

$$\text{dist}\big((x,\,y),\,(a,\,b)\big)=\sqrt{(x-a)^2+(y-b)^2} \qquad (4.3)$$

The nodes with minimum inter-node distance will form one cluster. Then, using the same formulae, we will calculate the distance of the node from the base station/sink. The node having minimum distance from the sink will be preferred as the cluster head in one criterion. Another criterion is the maximum amount of remaining energy. The node that has the maximum amount of remaining energy is more likely to become the cluster head. After the calculation of distances, some energy of nodes will be lost. So, the lost energy will be reduced from the total energy and the balanced energy will be obtained. The nodes' balance energy is the initial energy of the node minus the consumed energy of that node. The node with the maximum remaining energy and minimum distance from the sink node will become the cluster head. The cluster head node's energy will be tracked, and when it goes down the threshold, a new cluster head will be selected. The pseudocode for the MHBMO is given below:

```
Initialize worker
Set the positive properties.
Generate an population with positive properties
Set the best individual as the queen.
for i from 1 to max_mating_no_of_flights
        Queens_energy = random_energy()
        Queens_Speed = random_speed()
        while queens_energy > 0
                Queen_moves_between_states()
                Choose_drones_probabilistically()
                if (drone_selected = True) then
                        add_sperms_queens_spermathecal()
                endif
                update_queens_internal_energy()
                update_queens_speed()
        end while
        generate_broods_crossover_mutation_LMS()
        worker_improve_broods()
        if (best_brood = fitter_Queen) then
                replace_queen_with_best_brood()
```

```
      end if
      kill_all_broods()
end for
```

4.4.2 SIMULATION PARAMETERS

Some assumptions were made during the implementation of the proposed algorithm. The simulation parameters are shown in Table 4.1.

TABLE 4.1
Simulation Parameters

Sr. No.	Parameters	Value
1	Network area	400 m × 400 m
2	Initial energy	1 J
3	Packet size	100 bits
4	Transmission range	10 m
5	Transmission energy	50 nJ
6	Receiving energy	5 nJ
7	No. of nodes	1 – 400
8	No. of clusters	5, 10, 15, and 20

A total of n number of nodes is deployed in a square area of 400 m × 400 m. The whole area is divided into M × M dimensions. The energy model used the first-order radio model. Some assumptions have been made with respect to the sensor nodes in the deployment area and are as follows:

- The BS is placed outside of the deployment area, and its position is known to all the nodes.
- All the sensor nodes and BS are static.
- The nodes are in a homogenous environment and have the same initial energy E.
- The sensor nodes have the information regarding their energy and their location.

4.5 RESULTS AND DISCUSSION

The MHBMO algorithm is implemented. The results and its performance are analyzed here. Different cases are considered for the performance analysis of the existing HBMO and the modified HBMO. The results of the HBMO and MHBNO are recorded for varying number of nodes, i.e., from 100 to 400, and varying number of clusters from 5 to 20. On a sample basis, the results of the HBMO and MHBMO for 100 nodes and 5 clusters are presented here with the help of snaps for initialization of algorithm, network creation, cluster head selection, and balance energy in HBMO and MHBMO in Figures 4.3–4.7, respectively.

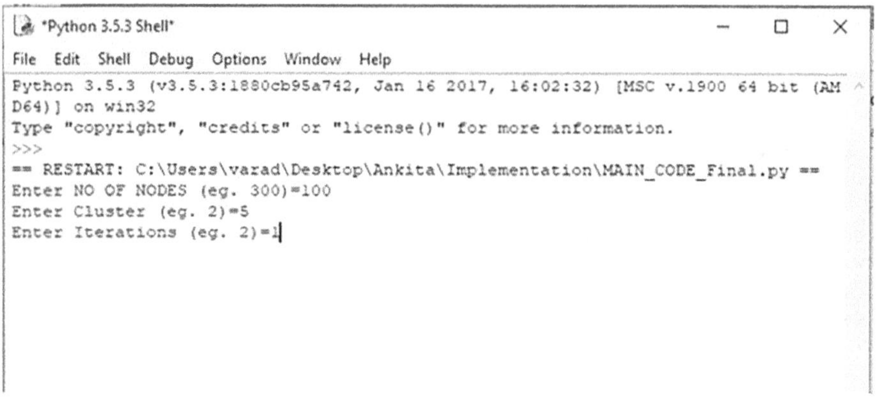

FIGURE 4.3 Initialization of algorithm.

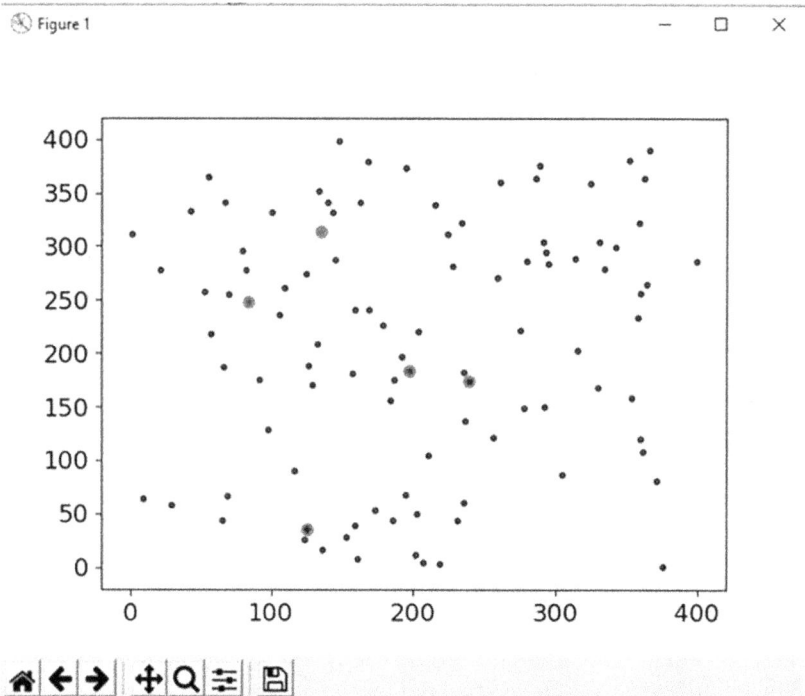

FIGURE 4.4 Network creation (100 nodes).

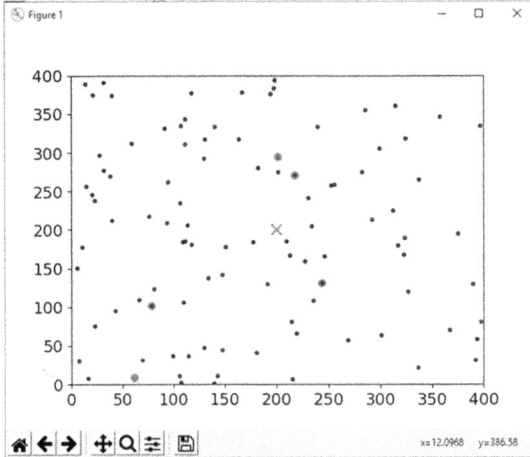

FIGURE 4.5 Cluster head selection.

```
xi (100, 100)
yi (100, 100)
x (100,)
y (100,)
z (100,)
z (100, 100)
PLOTT CONTOUR
x (100, 100)
y (100, 100)
z (100, 100)
ENTER TO CLUSTER LOOP
CONT---
x (100,)
y (100,)
z (100,)
xi (100, 100)
yi (100, 100)
x (100,)
y (100,)
z (100,)
z (100, 100)
PLOTT CONTOUR
x (100, 100)
y (100, 100)
z (100, 100)
{'DC': {'term2_initial': ('term2 learning', 'Rounds', [], 0, 1), 'term2_final':
('term2 final', 'Rounds', [], 0, 1), 'final_fitness': ('Final learning', 'Rounds
', [], 0, 1), 'term1_initial': ('term1 learning', 'Rounds', [], 0, 1), 'nb_sleep
ing': ('% of sleeping nodes', 'Rounds', [], 0, 1), 'coverage': ('Coverate rate',
 'Rounds', [], 0, 1), 'term1_final': ('term1 final', 'Rounds', [], 0, 1), 'overl
apping': ('Overlapping rate', 'Rounds', [], 0, 1), 'alive_nodes': ('Number of al
ive nodes', 'Rounds', [100], 1, 0), 'first_depletion': ('First depletion', 'Roun
ds', [0], 0, 0), '30per_depletion': ('30 percent depletion', 'Rounds', [0], 0, 0
), 'initial_fitness': ('Initial learning', 'Rounds', [], 0, 1)}}
()
Remaining energies:
[98.99999999999986]
[[0]]
HII
>>>
```

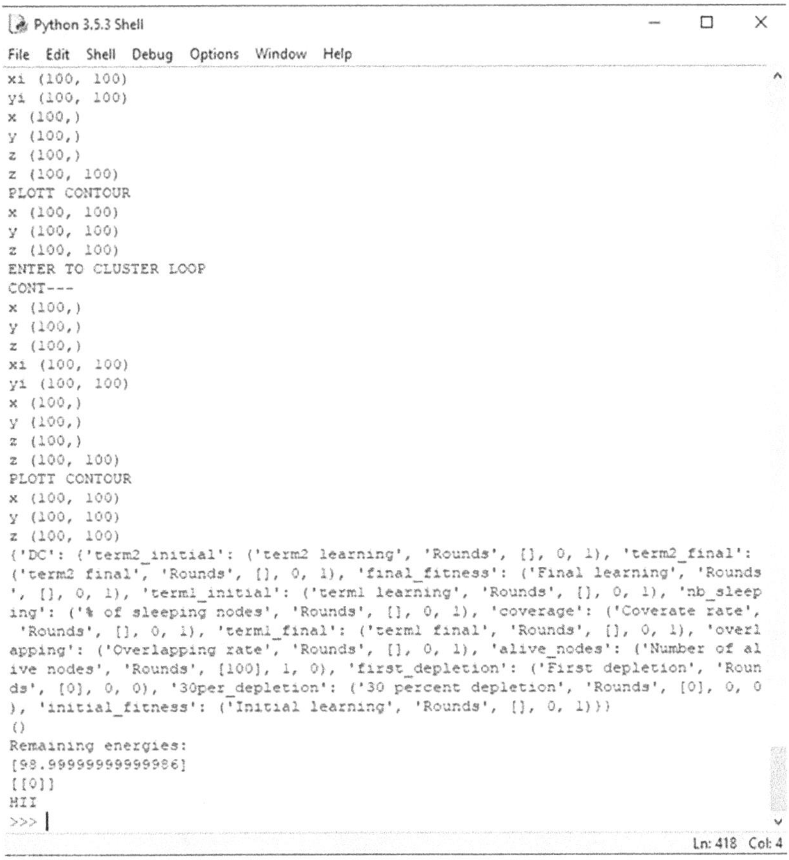

FIGURE 4.6 Balance energy using MHBMO.

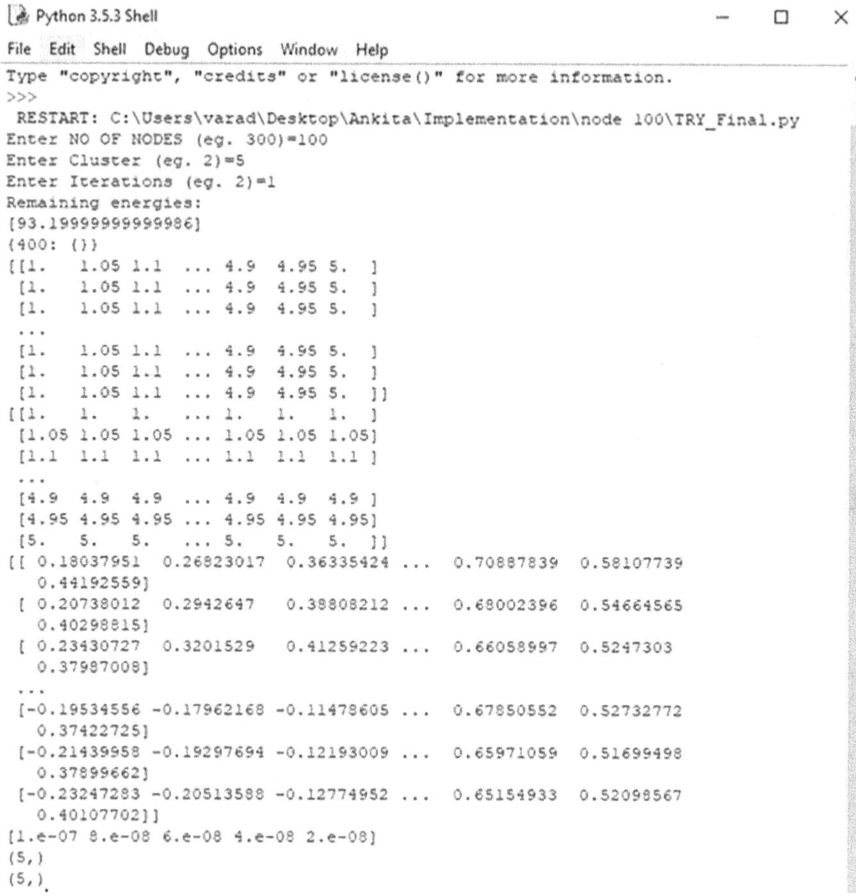

```
Python 3.5.3 Shell                                                       —    □    ×
File  Edit  Shell  Debug  Options  Window  Help
Type "copyright", "credits" or "license()" for more information.                    ^
>>>
 RESTART: C:\Users\varad\Desktop\Ankita\Implementation\node 100\TRY_Final.py
Enter NO OF NODES (eg. 300)=100
Enter Cluster (eg. 2)=5
Enter Iterations (eg. 2)=1
Remaining energies:
[93.19999999999986]
{400: {}}
[[1.    1.05 1.1  ...  4.9  4.95 5.  ]
 [1.    1.05 1.1  ...  4.9  4.95 5.  ]
 [1.    1.05 1.1  ...  4.9  4.95 5.  ]
 ...
 [1.    1.05 1.1  ...  4.9  4.95 5.  ]
 [1.    1.05 1.1  ...  4.9  4.95 5.  ]
 [1.    1.05 1.1  ...  4.9  4.95 5.  ]]
[[1.    1.    1.   ...  1.   1.   1.  ]
 [1.05 1.05 1.05 ...  1.05 1.05 1.05]
 [1.1  1.1  1.1  ...  1.1  1.1  1.1 ]
 ...
 [4.9  4.9  4.9  ...  4.9  4.9  4.9 ]
 [4.95 4.95 4.95 ...  4.95 4.95 4.95]
 [5.   5.   5.   ...  5.   5.   5.  ]]
[[ 0.18037951  0.26823017  0.36335424 ...  0.70887839  0.58107739
   0.44192559]
 [ 0.20738012  0.2942647   0.38808212 ...  0.68002396  0.54664565
   0.40298815]
 [ 0.23430727  0.3201529   0.41259223 ...  0.66058997  0.5247303
   0.37987008]
 ...
 [-0.19534556 -0.17962168 -0.11478605 ...  0.67850552  0.52732772
   0.37422725]
 [-0.21439958 -0.19297694 -0.12193009 ...  0.65971059  0.51699498
   0.37899662]
 [-0.23247283 -0.20513588 -0.12774952 ...  0.65154933  0.52098567
   0.40107702]]
[1.e-07 8.e-08 6.e-08 4.e-08 2.e-08]
(5,)
(5,)
```

FIGURE 4.7   Balance energy using HBMO.

Likewise, the results are recorded for 100, 200, 300, 400, and 500 nodes for 5, 10, and 20 clusters separately. The results are shown in Table 4.2.

By looking at this table, the implementation results obtained show that more energy is saved using MHBMO than the existing HBMO as the energy consumed using the existing HBMO is more than the MHBMO. The consumed energy and balance energy are calculated and recorded to calculate the energy saved. The performance analysis for energy saving using the HBMO and MHBMO is performed. Here, the initial energy allocated is 1 J/node. The graphical representation of performance analysis is shown. On a sample basis, the % energy consumption for 100, 200, 300, and 400 nodes with 5 clusters is shown in Figure 4.8 and the % balance energy for the same case is shown in Figure 4.9.

TABLE 4.2
Results of HBMO and MHBMO

No. of Clusters	No. of Nodes	Network Energy (J)	Results of HBMO			Results of MHBMO			% Energy Saving by MHBMO (J)
			Balance Energy (J)	Energy Consumed (J)	% Balance Energy (J)	Balance Energy (J)	Energy Consumed (J)	% Balance Energy (J)	
5	100	100	93.19	6.8	93.1	98.9	1.1	98.9	5.8
	200	200	187	13	93.5	198	2	99	5.5
	300	300	284	16	94.6	297	3	99	4.33
	400	400	381	19	95.2	396	4	99	3.75
10	100	100	92.2	7.8	92.2	98	2	98	5.8
	200	200	184.9	15	92.4	195.9	4.0	97.9	5.5
	300	300	280.9	19	93.6	293.9	6.0	97.9	4.33
	400	400	377	23	94.2	392	8.0	98	3.75
15	100	100	91.1	8.8	91.1	96.9	3.0	96.9	5.8
	200	200	182.9	17	91.4	193.9	6.0	96.9	5.5
	300	300	278	22	92.6	291	9.0	97	4.33
	400	400	373	27	93.2	388	12	97	3.75
20	100	100	90.1	9.8	90.1	95.9	4.0	95.9	5.8
	200	200	181	19	90.5	192	8.0	96	5.5
	300	300	275	25	91.6	288	12	96	4.33
	400	400	368.9	31	92.2	383.9	16	95.9	3.75

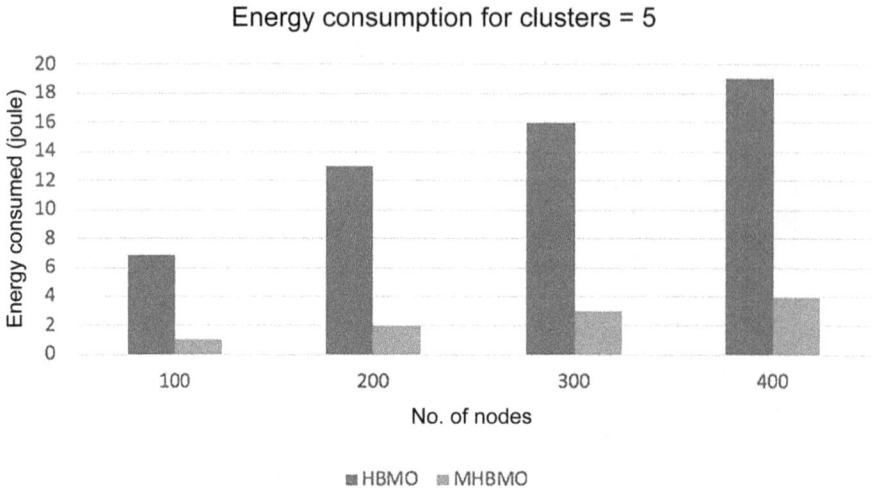

FIGURE 4.8 Percentage energy consumed for 100, 200, 300, and 400 nodes with 5 clusters.

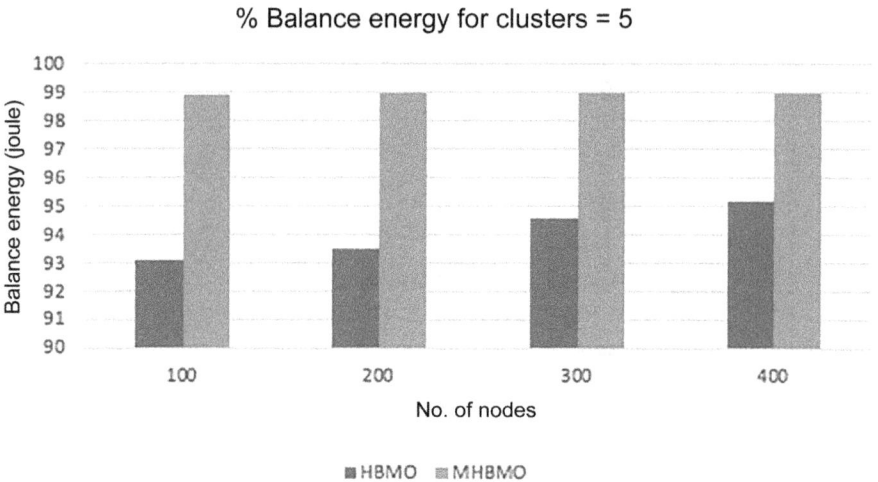

FIGURE 4.9 Percentage balance energy for 100, 200, 300, and 400 nodes with 5 clusters.

Likewise, the energy consumed, balance energy, and saved energy when using the HBMO and MHBMO for 100, 200, 300, and 400 nodes with 10, 15, and 20 clusters are also calculated and recorded. The comparative % energy saving using MHBMO with that of HBMO for 5 clusters with 100, 200, 300, and 400 nodes is shown in Figure 4.10.

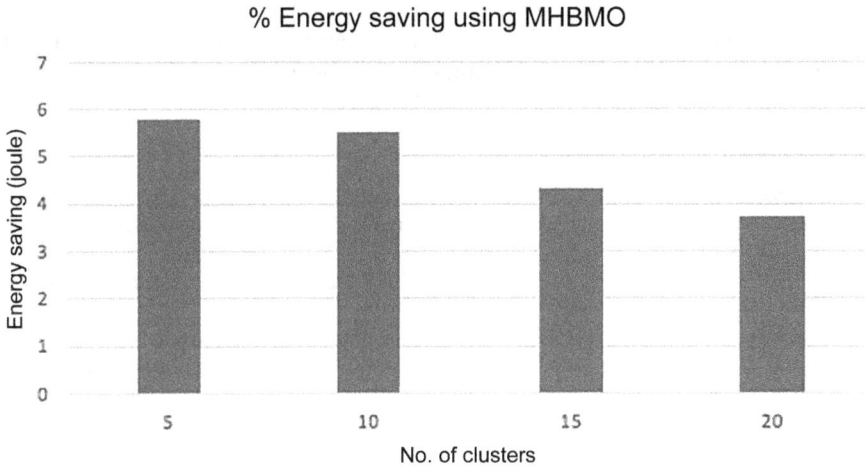

FIGURE 4.10 Percentage energy saving when using 5, 10, 15, and 20 clusters.

4.6 CONCLUSION

The modified HBMO algorithm selected nodes for forming clusters based on positive properties, such as the minimum distance between two nodes, minimum distance of the node from the sink, and maximum amount of remaining energy, instead of randomly selecting nodes. Then, LMS was applied for selecting the most appropriate nodes from the eligible nodes list to form the energy-efficient cluster. The comparative analysis showed that the proposed algorithm consumes less amount of energy compared to the existing algorithm and extends the network's life. This chapter offered an effective cluster head selection algorithm based on HBO algorithm. This algorithm considered the parameters of energy and distance to improve the selection of the cluster head. The main goal of EDC-HBO was to enhance the network's life as well as to minimize energy consumption. The simulation results showed that EDC-HBO is more energy efficient than LEACH and UCR. This system fulfilled all the objectives that make the system more efficient than the existing system. The energy saved using the proposed algorithm, i.e., MHBMO, was 98.99%, whereas the existing algorithm saved 93.19%. So, the percentage energy analysis comes out to be as 5.8%. Similarly, for all the cases, the results are positive. For 200 nodes, the average percentage energy analysis comes out to be 5.5%, for 300 nodes, it results in 4.333%, and for 400 nodes, the output obtained is 3.75%. In all the cases, the modified HBMO algorithm gave better results.

ACKNOWLEDGMENT

We thank our PG student Ms. Ankita Bhusari who helped us with the implementation. We also acknowledge the support provided by our colleagues from SCET, MIT

AOE, Alandi (D.), Pune, who provided the suggestions and guidance to reach up to a considerable level. We thank Dr. Barani S. and Dr. Sanjeev J. Wagh for their assistance and comments that greatly improved this work.

REFERENCES

1. Jafarizadeh, V., A. Keshavarzi, and T. Derikvand. Efficient cluster head selection using Naïve Bayes classifier for wireless sensor networks. *Wireless Networks* 23.3 (2017): 779–785.

2. Zahedi, A., et al. Energy efficient reservation-based cluster head selection in WSNs. *Wireless Personal Communications* 100.3 (2018): 667–679.

3. Vijayalakshmi, K., and P. Anandan. A multi objective Tabu particle swarm optimization for effective cluster head selection in WSN. *Cluster Computing* 22.5 (2019): 12275–12282.

4. Selvi, M., et al. HBO based clustering and energy optimized routing algorithm for WSN. *2016 Nineth International Conference on Advanced Computing (ICoAC)*, Chennai, India. IEEE, 2017.

5. Sengottuvelan, P., and N. Prasath. BAFSA: Breeding artificial fish swarm algorithm for optimal cluster head selection in wireless sensor networks. *Wireless Personal Communications* 94.4 (2017): 1979–1991.

6. Daflapurkar, P.M., M. Gandhi, and B. Patil. Tree based distributed clustering routing scheme for energy efficiency in wireless sensor networks. *2017 IEEE International Conference on Power, Control, Signals and Instrumentation Engineering (ICPCSI)*, Chennai, India. IEEE, 2017.

7. Jha, S.Kr., and E.M. Eyong. An energy optimization in wireless sensor networks by using genetic algorithm. *Telecommunication Systems* 67.1 (2018): 113–121.

8. Murugan, T.S., and A. Sarkar. Optimal cluster head selection by hybridisation of firefly and grey wolf optimisation. *International Journal of Wireless and Mobile Computing* 14.3 (2018): 296–305.

9. Ferdous, R., V. Muthukkumarasamy, and E. Sithirasenan. Trust-based cluster head selection algorithm for mobile ad hoc networks. *2011 IEEE 10th International Conference on Trust, Security and Privacy in Computing and Communications*, Changsha, China. IEEE, 2011.

10. Handy, M.J., M. Haase, and D. Timmermann. Low energy adaptive clustering hierarchy with deterministic cluster-head selection. *4th International Workshop on Mobile and Wireless Communications Network, Stockholm, Sweden.* IEEE, 2002.

11. Gupta, G., and M. Younis. Load-balanced clustering of wireless sensor networks. *IEEE International Conference on Communications, ICC 2003*, Anchorage, AK, USA. Vol. 3. IEEE, 2003.

12. Nazir, B., and H. Hasbullah. Energy balanced clustering in wireless sensor network. *2010 International Symposium on Information Technology*, Kuala Lumpur, Malaysia. Vol. 2. IEEE, 2010.

5 Multiobjective Design Optimization of Power Take-Off (PTO) Gear Box Through NSGA II

R. Saravanan
Vaigai college of Engineering

G. Chandrasekaran
Sree Sakthi Engineering College

V. S. Sree Balaji
Roever Engineering College

CONTENTS

5.1 INTRODUCTION

A power take-off (PTO) is one of the several methods used for taking power from a power source, such as a running engine, and transmitting it to an application such as an attached implement or separate machine. A typical PTO is designed to pick up power from the engine, through rotation, and transfer it to another piece of equipment.

There is a need for lightweight gear box designs with a high load capacity and power density with high reliability of the gears, taking resources into consideration [1]. All these conditions result in special design criteria with the consideration of maximum load, in addition to dynamic loads under different load situations. Hence, a procedure to automate the optimization of the design of individual pairs and gear trains, considering several objective functions and constraints, is being attempted. In this situation, traditional algorithms are not sufficient and there is a need for an efficient algorithm for accomplishing this task.

Hence, the problem is identified and the experimental investigation of an existing production PTO gear box consisting of an epicyclic gear train is carried out. NSGA II is implemented to find the optimized design solution.

5.2 MATHEMATICAL FORMULATION OF MULTIOBJECTIVE PROBLEMS

Multiobjective evolutionary algorithms simulate the evolution of individual structures via the processes of selection and reproduction [2]. For computational design optimization, objective function and constraints must be expressed as a function of design variables (or design vector X):

Objective function: $f(x)$
Constraints: $g(x)$, $h(x)$
Optimization statement: Minimize $f(x)$,
subject to $g(x) \leq 0$

$$h(x) = 0$$

where
   $f(x)$: the objective function to be minimized.
   $g(x)$: the inequality constraints.
   $h(x)$: the equality constraints.
   x: the design variables.

5.3 NON-DOMINATED SORTING GENETIC ALGORITHM – NSGA II

By using the multiobjective genetic algorithm (MoGA) optimization technique, problems with more than one objective function are concurrently optimized [3]. Several Pareto optimal points can be simultaneously obtained with an even distribution of solutions from the Pareto optimal sets, which can reduce the production time and cost [4]. There are many new multiobjective optimization techniques that are enhancements of single-objective GA optimization, and the literature shows that among all MoGA techniques, NSGA II has been the most widely applied for optimizing gear design parameters [5]. The interest of this study is to review the application of the NSGA II in optimizing gear design parameters.

The NSGA II represents the state-of-the-art of multiobjective evolutionary optimization. Proposed by K. Deb [6] on the basis of NSGA in 2002, NSGA II improves the non-dominated sorting algorithm and reduces the computational complexity. It sorts the combination of parents and children population with an elitist strategy, introduces the crowded comparison operator to improve the diversity of solutions, and avoids the use of niched operators.

5.4 PROBLEM STATEMENT OF PTO GEAR BOX DESIGN OPTIMIZATION

The procedure reported herein provides an optimization of the basic type of epicyclic gear train of planetary type. The presented approach is based on suitably modifying the original NSGA II algorithm, which is the basis of the mathematical model. There is a need for lightweight construction of gear box designs with a high load capacity, power density, and reliability of the gears, taking resources into consideration.

All these conditions result in special design criteria with the consideration of maximum of load, in addition to dynamic loads under different load situations. Hence, a procedure to automate the optimization of the design of individual pairs, considering several objective functions and constraints, is being attempted.

The optimization of the planetary gear transmission is carried out using three objective criteria such as minimization of weight, maximization of efficiency, and minimization of center distance between gear pairs. The application of multiobjective optimization methods for choosing an optimal solution from Pareto front is carried out in this work. The aim is to find a set of Pareto optimal solutions in a single run.

5.4.1 CASE STUDY

An existing PTO gear box having a spiral bevel gear reduction stage and a planetary gear reduction stage is taken for the case study. The planetary gear train consists of a sun gear, which is driven by the previous spiral bevel gear reduction unit, five planet gears mounted on the spider arm, and a stationary ring gear. To find the speed of individual elements of the existing planetary gear train, let us consider the following:

No. of teeth on planet gears $P = t_p = 15(z_1)$
No. of teeth on the annular ring gear $A = t_a = 56(z_2)$
No. of teeth on the sun wheel $S = t_s = 24(z_3)$
The spider arm is L.

The last row of Table 5.1 provides us the equation for the speed of individual elements.

Input speed of bevel pinion = 500 rpm
Speed of bevel gear $= \dfrac{500}{4.778} = 104.651$ rpm
Speed of sun wheel = Speed of bevel gear = 104.651 rpm
Speed of planets $= y - \dfrac{24}{15}x = 31.3953 - \dfrac{24}{15} \times 73.2557 = -85.814$ rpm (ACW)

TABLE 5.1

Speed Calculations for Planetary Elements

Operation	Spider Arm L	Sun Wheel t_s	Planet Wheel t_p	Annular Wheel t_a
1. Arm L is locked and turn wheel S is given +1 revolution	0	+1	$-\dfrac{t_s}{t_p} = -\dfrac{24}{15}$	$-\dfrac{t_s}{t_a} = -\dfrac{24}{56}$
2. Multiply by x	0	x	$-\dfrac{t_s}{t_p}x = -\dfrac{24}{15}x$	$-\dfrac{t_s}{t_a}x = -\dfrac{24}{56}x$
3. Add y to all columns	0	$y+x$	$y-\dfrac{t_s}{t_p}x = y-\dfrac{24}{15}x$	$y-\dfrac{t_s}{t_a}x = y-\dfrac{24}{56}x$

When the annulus is locked, $y - \dfrac{24}{56}x = 0$

$$y + x = 104.651$$

$$y = 104.651 - x$$

Substituting (3) in (1), $104.651 - x - \dfrac{24}{56}x = 0$

$$x\left(1+\dfrac{24}{56}\right) = 104.651$$

$$\therefore x = 73.2557$$

Speed of the spider arm $y = 31.3953\,\text{rpm}\,(\text{CW})$

Bevel reduction $i_b = \dfrac{43}{9} = 4.778$

Epicyclic (planetary) reduction $i_p = 3.3$ [7]

Total reduction $i = i_b \times i_p = 4.778 \times 3.333 = 15.926$

5.4.2 OBJECTIVE FUNCTIONS AND CONSTRAINTS

Three objective functions for minimization/maximization with an array of inequality and equality constraints have been considered, and the problem can be stated as follows.

The following objective functions are considered for the design optimization of the planetary gear train:

$$F_{\text{obj}} = \begin{cases} \text{Min. center distance between planetary gear pairs} \\ \text{Min. weight of the planetary gear train} \\ \text{Max. efficiency of planetary gear train} \end{cases}$$

The independent design variables of the spiral planetary gear train include the module and number of teeth of pinion. The vector of design variables is given below:

$$X(i) = \begin{cases} \text{module } m \\ \text{No. of pinion teeth } z_1 \\ \text{width-to-module ratio } \psi_m \\ \text{No. of planets } n_w \end{cases}$$

where $i = 1,\dots$, number of variables.

A number of constraints, which act as sub-functions to restrict the objective function, are introduced to incorporate suitable contents. The following constraints are considered for the design optimization of the planetary gear train:

$$g(j) = \begin{cases} \text{Bending stress constraint} \\ \text{Surface contact compressive stress constraint} \\ F_S \geq F_d \text{ constraint} \\ F_w \geq F_d \text{ constraint} \\ \text{Surface pressure constraint} \\ \text{Scoring index constraint} \\ \text{Involute interference of internal gearing constraint} \\ \text{Contact ratio of internal gear pair constraint} \\ \text{Tip interference of internal gearing constraint} \\ \text{Condition for proper assembly of planetary gearing constraint} \end{cases}$$

where $j = 1, \dots$, number of constraints.

Since the center distance and weight should be minimized and the efficiency should be maximized, the following notation is denoted in this model, since NSGA II is essentially a minimization software:

$$f_1 = a$$
$$f_2 = W$$
$$f_3 = -\eta$$

5.4.3 DESIGN VARIABLES

The independent design parameters of the planetary gear drive include module m, the number of teeth of pinion z_1, width-to-module ratio ψ_m, and the number of planets n_w. So the design variables are as follows:

$$X = \begin{bmatrix} m_t, z_1, \psi_m, n_w \end{bmatrix}^T = \begin{bmatrix} x_1, x_2, x_3, x_4 \end{bmatrix}^T$$

TABLE 5.2

Planetary Gear Parameters

Data	Gear
Density of the material ρ	7.86×106 kg/mm³
Material of the sun gear and planet gear	Steel 20Mn5Cr5 (IS:4432-1988)
Design bending stress $[\sigma_b]$	$373\,N/mm^2$
Design crushing stress $[\sigma_c]$	$1100\,N/mm^2$
Input speed N	104.65 rpm
Power transmitted P	50 HP
Young's modulus of the material (E)	$2.0 \times 10^5\,kg/mm^2$
Pressure angle ϕ_n	25°
Ratio between face width and average module ψ_m	13.225

5.5 PROBLEM FORMULATION FOR OPTIMIZATION

This section describes the problem formulation for the existing production planetary gear for the optimization of weight, efficiency, and center distance.

The parameters considered for the planetary gear design based on the existing gear design are given in Table 5.2.

5.5.1 PLANETARY GEAR DESIGN OPTIMIZATION FORMULATION

The complete planetary gear design problem is as formulated as below:

1. Minimize center distance $f_1 = \min a = 1.3667\,mz_1$

2. Minimize total weight
$$f_2 = W = \rho\psi_m m^3 \begin{bmatrix} 9.577(z_1 - 3.05)n_w + \\ 12.56(3.733z_1 + 4) + \\ 6.28(1.6z_1 - 2) \end{bmatrix}$$

3. Maximize efficiency [8] $f_3 = \eta = 100 - (P_{L1} + P_{L2})0.7$

where

$$P_{L1} = \frac{50f}{\cos\phi}\left(\frac{H_{s1}^2 + H_{t1}^2}{H_{s1} + H_{t1}}\right)$$

$$P_{L2} = \frac{50f}{\cos\phi}\left(\frac{H_{s2}^2 + H_{s2}^2}{H_{s2} + H_{t2}}\right)$$

$$H_{s1} = 1.625\left[\sqrt{\left(1 + \frac{3.2}{z_1}\right)^2 - 0.8214} - 0.42262\right]$$

$$H_{t1} = 2.59898 \left[\sqrt{\left(1 + \frac{2}{z_1}\right)^2 - 0.8214} - 0.42262 \right]$$

$$H_{s2} = 2.733 \left[-\sqrt{\left(1 - \frac{1.07158}{z_1}\right)^2 - 0.8214} + 0.42262 \right]$$

$$H_{t2} = 0.73212 \left[\sqrt{\left(1 + \frac{3.2}{z_1}\right)^2 - 0.8214} - 0.42262 \right]$$

subject to

1. $\sigma_b = \dfrac{20590935.65}{n_w m^3 \psi_m z_1} \leq 373$

2. $\sigma_c = \dfrac{1062682.7}{m z_1} \sqrt{\dfrac{1}{n_w m \psi_m}} \leq 1100$

3. $F_b \geq F_d$
4. $F_w \geq F_d$
5. $\sigma_{NT} \leq 2300$ [9]
6. $SI_m \leq 135$ [10]
7. $0.04938 z_1 \geq 1$ [11]
8. $2.145 \tan\left(\cos^{-1} X\right) + 0.2679 \geq 1$ [11]

   where $X = \left(\dfrac{2.1147 z_1}{2.333 z_1 - 2} \right)$

9. $m_p = 0.1756 \left[\begin{array}{l} \left(0.06977 z_1^2 + 2.5 z_1 + 4\right)^{\frac{1}{2}} + 0.72188 z_1 \\ -\left(0.9707 z_1^2 - 9.33325 z_1 + 4\right)^{\frac{1}{2}} \end{array} \right] \geq 1.4$ [12]

10. $x_2 - x_1 \geq 0$ [11,13]

11. $\dfrac{z_2 + z_3}{n_w} = I$, where I is an integer [8].

5.5.2 Variable Bounds

Transverse module and the number of pinion teeth are considered as variables (x_1, x_2), and their lower and upper bounds considered are as follows:

Module m: 1–6 mm.
No. of pinion teeth: 10–23.
Ψ_m: 10–23 mm.
n_w: 2–6 mm.

5.5.3 Input Parameters

The objective functions and design constraints formulated as in Section 5.5.1 are entered in the software coding, making necessary modifications as required for the optimization. The probability of crossover and mutation are taken as per the recommendations given in the software.

5.6 RESULTS AND DISCUSSION

In this section, the results obtained by NSGA II are discussed and compared with the design of the existing production gear pair.

In Table 5.3, a sample of NSGA II output is shown. Since NSGA II software accepts constraints of the form of $\geq$ only, all the columns corresponding to the four constraints shown in Table 5.3 indicate the results obtained by modifying the constraint equations accordingly.

In Table 5.4, a comparison is made between the results that are obtained with NSGA II and the design of the existing production gear pair for the best performance. The number of teeth of pinion was selected as 23 from the best populations of NSGA II results for comparison. The values of variables and objective functions are indicated in Table 5.4 for comparison.

From the NSGA II results, the nearest integer values of the number of teeth of pinion Z_1 of 23 were taken for comparison with the existing gear pair. The total weight of pinion and gear in kg, the efficiency, and the center distance in mm were taken for comparison, as shown in Table 5.4.

The comparison shows that the results obtained from NSGA II software are superior by optimizing the three objective functions without any constraint violation. The results were obtained with minimum computational effort and time.

Figure 5.1 shows the 3D plot of the objective functions, viz. total weight, efficiency, and cone distance of the best population generated by NSGA II. Here, all generations were having no constraint violation. The Pareto front is formed for the population of the best solutions.

Moving along the frontier, from left to right, we improve the value of all the objective functions, viz. f_1, f_2, and f_3. Pareto solutions located close to the upper right corner have a higher total weight, a higher center distance, and a higher efficiency, which means they have better weight and efficiency but lesser compactness. On the contrary, the solutions near the lower left corner present better compactness and better total weight but lesser efficiency. As a result of optimization, a Pareto optimal front is formed in the objective space of the solutions. The designers are able to make the most proper choice that satisfies their practical requirements. It can also be observed that the distribution of solutions along the front is uniform, which means the NSGA II algorithm offers a high-quality solution set.

Figure 5.2 shows the comparison of the objective functions for the existing spiral gear pair design and the optimized design based on NSGA II results. The design with the number of teeth on pinion as 23 from NSGA II shows the best optimal reduction of 17.43% in weight with improved efficiency of 0.1% compared to the existing production gear set.

TABLE 5.3
Sample of NSGA II Output (Planetary Gear Train Design)

No. of Objectives = 3 | No. of Constraints = 10 | No. of Real Var = 4

Center Distance (mm)	Weight (kg)	Efficiency	Bending Stress	Crushing Stress	$F_b > F_d$	$F_w > F_d$	Hertz Stress	Scoring Index	Involute Interference 1	Involute Interference 2	Contact Ratio	Tip Interference	Module (mm)	No. of Pinion Teeth	ψ_m	n^w
83.2	8.1	99.4	0.0	155.2	40.9	11982.9	1796.5	37.3	0.8	0.1	0.2	0.0	2.7	22.9	22.6	5.7
68.7	8.9	99.1	22.2	2.9	5.9	7266.7	1621.4	34.1	0.3	0.1	0.2	0.0	3.1	16.0	23.0	5.1
84.5	8.1	99.4	1.7	159.1	165.3	12087.1	1799.2	36.5	0.8	0.1	0.2	0.0	2.7	23.0	21.9	5.7
69.0	9.0	99.1	30.1	17.6	180.3	7515.8	1626.8	35.7	0.3	0.1	0.2	0.0	3.1	16.0	23.0	5.2
78.6	8.5	99.3	17.2	123.2	338.4	10701.5	1759.0	38.0	0.6	0.1	0.2	0.0	2.8	20.4	22.6	5.6
70.1	8.8	99.1	15.9	13.6	72.5	7737.3	1649.2	33.6	0.3	0.1	0.2	0.0	3.1	16.6	22.9	5.1
69.9	8.9	99.1	18.8	10.4	237.2	7767.9	1641.7	33.4	0.3	0.1	0.2	0.0	3.1	16.4	22.8	5.1
70.4	8.7	99.2	10.1	10.1	127.5	7918.2	1656.4	32.6	0.3	0.1	0.2	0.0	3.1	16.8	23.0	5.1
73.0	8.6	99.2	1.1	27.9	631.2	9206.7	1696.9	31.4	0.4	0.1	0.2	0.0	3.0	17.7	22.9	5.0
70.8	8.7	99.2	10.5	13.1	36.9	7815.1	1658.0	32.4	0.3	0.1	0.2	0.0	3.1	16.8	22.6	5.1

TABLE 5.4

Optimized Design Variables and Objective Functions of Planetary Gear Train Using NSGA II

	Variables		Objective Functions		
	Module m (mm)	No. of Teeth (Planet Gear) Z_1	Total Weight W (kg)	Efficiency η	Center Distance (mm)
Existing gear pair design	4	15	9.81	99.3	82
NSGA II design	2.7	23	8.1	99.4	83.2

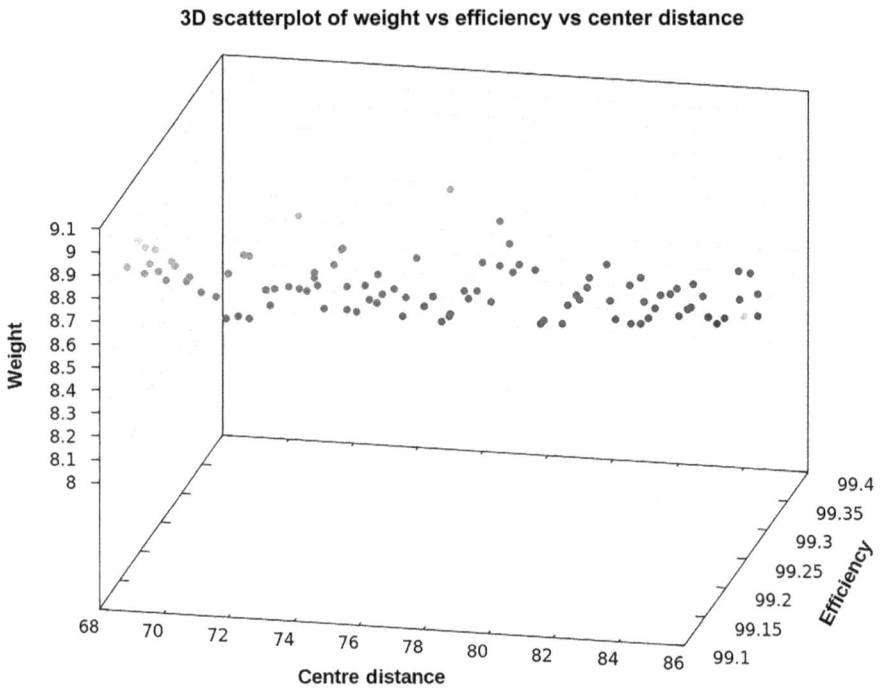

FIGURE 5.1   Pareto front of the planetary gear train design – three objective functions.

5.6.1 CONDITION FOR PROPER ASSEMBLY

Applying the condition for proper assembly for the NSGA II results, we can obtain the number of teeth for the ring gear and the sun gear.

That is, for $Z_1 = 23$,

Planet speed $N_1 = 31.392 - \dfrac{Z_3}{Z_1} 73.2557$; initially assuming the same value for x,

$$Z_3 = 36.8$$

■ Existing gear pair ■ NSGA II – pinion no. of teeth = 23

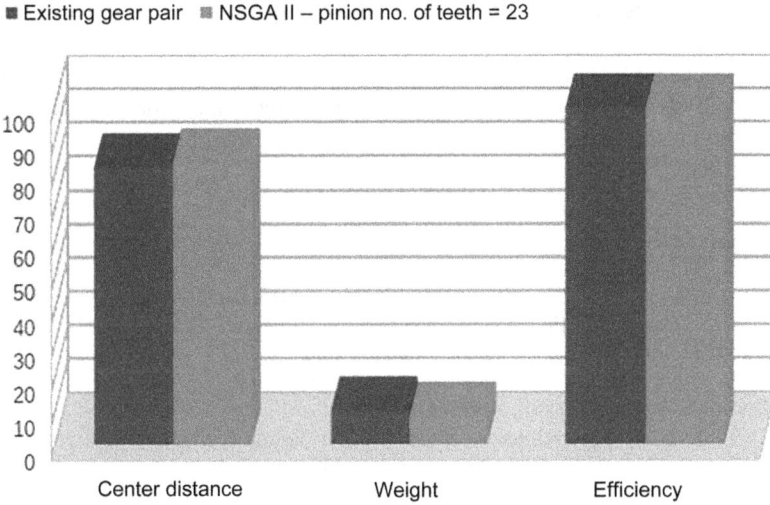

FIGURE 5.2 Comparison of the existing and optimized NSGA II designs (planetary gear).

Since the ring gear is fixed,

i.e., $y - \dfrac{Z_3}{Z_2} x = 0$, and initially assuming $y = 31.3953$ as before,

$$Z_2 = 86.33$$

Applying the condition for proper assembly,

$\dfrac{Z_2 + Z_3}{n_w} = I$, i.e., taking $Z_3 = 34$ and $Z_2 = 86$ satisfies this condition, for the total number of planets $n_w = 6$.

Hence, $x = \dfrac{104.65}{1 + \dfrac{Z_3}{Z_2}} = 75$ and $y = 104.65 - 75 = 29.65\,\text{rpm}$

The total reduction then is $\dfrac{500}{29.65} = 16.86$, which is slightly more than the total reduction 15.926 of the existing design and hence is acceptable.

5.7 CONCLUSIONS

In this chapter, a reliable model was developed for the multiobjective design optimization of planetary gear trains. The non-dominated sorting genetic algorithm
(NSGA II) procedure was applied having three objective functions with necessary
constraints for design optimization. The general equations for the objective functions
and the constraints were formulated in terms of four variables, viz. module, the number of teeth on pinion, width-to-module ratio (Ψ_m), and the number of planets. The
NSGA II algorithm was modified as necessary, to suit the present requirements, and
then compiled and run, supplying the necessary input data.

Supported by the in-depth results and analytical tables, the planned work achieved better results when compared to other existing methods. This technique helps to find the Pareto optimal front that contains the best possible optimal solution, from which the required solution can be obtained. By using this technique, the designers are able to make the most proper choice that satisfies their practical requirements, at a lesser cost and time.

REFERENCES

1. T. S. G. Souza, D. M. M. de Souza, and J. Savoy, 2013, Light-Weight Assembled Gears: A Green design Solution for Passenger and Commercial Vehicles, Gear Technology, May 2013.
2. D. E. Goldberg, 1989, *Genetic Algorithms in Search, Optimization and Machine Learning*, Addison-Wesley, Reading, MA.
3. K. Deb, 2011, Multi-Objective Optimization Using Evolutionary Algorithms: An Introduction, KanGAL Report Number 2011003.
4. K. Deb, S. Chaudhuri, and K. Miettinen, 2005, Estimating Nadir Objective Vector Quickly Using Evolutionary Approaches, KanGAL Report Number 2005009.
5. R. Datta and K. Deb, 2010, A classical-cum-evolutionary multi-objective optimization for optimal machining parameters, pp. 607–612. doi:10.1109/NABIC.2009.5393425
6. K. Deb, A. Pratap, S. Agarwal, and T. Meyarivan, 2002, A fast and elitist multiobjective genetic algorithm: NSGA-II, *IEEE Transactions on Evolutionary Computation*, Vol. 6, No. 2, 182–197.
7. W. D. Darle, 1962, *Gear Handbook: The Design, Manufacture and Application of Gears*, McGraw-Hill Book Co., Denver, CO.
8. D. P. Townsend, 2013, *Dudley's Gear Handbook - The Design, Manufacture and Applications of Gears*, McGraw Hill, New York.
9. M. Savage, J. J. Coy, and D. P. Townsend, 1982, Optimal tooth numbers for compact standard spur gear sets, *Journal of Mechanical Design*, Vol. 104, 749.
10. M. Hirt, 1984, An improved method to determine the scoring resistance of high power high speed gearing, *Proceedings of the Thirteenth Turbo Machinery Symposium*, College Station, TX.
11. E. Buckingham, 1988, *Analytical Mechanics of Gears*, Dover Publications, New York.
12. R. Ananthapadmanabhan, 2016, Investigation on multiple algorithms for multi-objective optimization of gear box, *IOP Conference Series: Materials Science and Engineering*, Rourkela, India.
13. S. A. E. Andersson, 1973, *On the Design of Internal Involute Spur Gears, Transactions of Machine Elements Division*, Lund Technical University, Lund, Sweden.

6 Improving the Performance of Machining Processes Using Opposition-Based Learning Civilized Swarm Optimization

P. J. Pawar and K. N. Nandurkar
Department of Production Engineering, K.K. Wagh
Institute of Engineering Education and Research

CONTENTS

6.1 INTRODUCTION

Every optimization method tries to obtain the utmost benefit from the available resources. Most of the traditional methods of optimization such as descent methods (Cauchy method, Marquardt method, quasi-Newton methods, etc.), Lagrange multiplier method, generalized reduced gradient, and sequential quadratic programming are gradient based. Although gradient-based methods have proved their capability to solve many complex real-life problems, there is a high possibility that these methods might stuck in a local minima/maxima when traversing parameter(s), especially in case of multimodal problem. This is due to the fact that the gradient at any local minima/maxima is zero, thus making the difference between the current and the next parameter also zero. Since most of the real-life mathematical models (either explicit or implicit) are highly multimodal, gradient-free probabilistic approaches that overcome this limitation due to their capability to climb the hill are quite useful. If appropriate trade-off between exploitation and exploration is provided, these gradient-free approaches are highly successful to find the global optimum solution. Examples of gradient-free approaches include evolutionary algorithms, which evolve either genetically (viz. genetic algorithms, memetic algorithms, etc.) or by their social behaviour (viz. particle swarm optimization, shuffled frog leaping algorithm, ant colony optimization algorithm, firefly algorithm, etc.), and other metaheuristics such as simulated annealing and tabu search.

Evolutionary algorithms are population based, in which a set of initial solutions (population) is generated and updated iteratively. Each new generation is produced by stochastically removing less desired solutions, and introducing small random changes. Although these algorithms are very efficient to find the global optimum solution, the main challenge for the user is to tune appropriately their control parameters to suit to specific application. The genetic algorithm proposed by Holland (1975) may be considered as the first most popular metaheuristic algorithm. However, it suffers a major drawback of tuning a large number of algorithm-specific parameters (such as crossover probability, mutation probability, selection operator, etc.). Particle swarm optimization (PSO) proposed by Kennedy and Eberhart (1995) overcomes this limitation as it requires only three control parameters (inertia weight, cognitive parameter, and social parameter) to tune. However, the limitations of this algorithm are that it can easily fall into a local optimum in a high-dimensional space and has a low convergence rate in the iterative process. Civilized swarm optimization proposed by Selvakumar and Thanushkodi (2009) is an improved version of particle swarm optimization, which overcomes these limitations of particle swarm optimization algorithm by combining the benefits of particle swarm optimization and society–civilization algorithm (Ray and Liew, 2003). Since particle swarm optimization and society civilization algorithm provide the basis for civilized swarm optimization algorithm, this chapter provides brief information about these two contributing algorithms. The steps of civilized swarm optimization are then discussed in detail.

Since evolutionary algorithms start searching with a randomly generated initial population of solutions, it is possible that these solutions may not have the capability to explore the complete search space. This may lead to premature convergence, a large number of generations to converge, or getting trapped into a local optimal solution.

To overcome this drawback, a technique of opposition-based learning introduced by Tizhoosh (2005) is presented in this chapter. We can improve our chance to start with a closer (fitter) solution by checking the opposite solution simultaneously. The purpose of this study is to demonstrate the effectiveness of opposition-based learning civilized swarm optimization for parametric optimization of machining processes.

6.2 METHODOLOGY

Since civilized swarm optimization (CSO) algorithm is an integrated technique of particle swarm optimization (PSO) and society civilization algorithm (SCA), these two algorithms are briefly discussed in this section. The CSO algorithm is then explained in a step-by-step manner through a simple example. The concept of opposition-based learning is demonstrated at the end of this section.

6.2.1 PARTICLE SWARM OPTIMIZATION

Particle swarm optimization (PSO) is an evolutionary computational technique developed by Kennedy and Eberhart (1995). This algorithm models the intelligent behaviour of a flock of migrating birds. Every bird starts its journey from a different location and arrives at the same particular location at the same time. The birds interact between members of swarm to share information. Based on this information, every bird updates its velocity and position using Eqs. (6.1) and (6.2), respectively.

$$V_{i+1} = w \times V_i + c_1 \times r_1 \times \left(p\text{Best}_i - X_i \right) + c_2 \times r_2 \times \left(g\text{Best}_i - X_i \right) \tag{6.1}$$

$$X_{i+1} = X_i + V_{i+1} \tag{6.2}$$

As given by Eq. (6.1), the bird updates its velocity by using the information about the current velocity (V_i), its own best position achieved so far ($p\text{Best}$), and the global best position ($g\text{Best}$). w, c_1, and c_2 are algorithm-specific parameters representing inertia weight, confidence of bird in itself, and confidence of bird in swarm. r_1 and r_2 are the independently generated random numbers between 0 and 1. These random numbers represent the ever-changing nature of self-confidence (c_1) and confidence in swarm (c_2). Once the velocity is updated, the bird changes its position according to Eq. (6.2). The values of inertia weight (w) and acceleration constants (c_1 and c_2) should be tuned appropriately to ensure the convergence of the algorithm for a given problem. As mentioned by the originators, the value of inertia weight (w) should be between 0 and 1 and the values of acceleration coefficients should be between 0 and 2. Few guidelines about the optimum selection of three parameters based on convergence behaviour of PSO algorithms were given by Bergh and Engelbrecht (2006). An important feature of PSO algorithm is that at convergence, all solutions are the same, i.e. the best solution. However, in PSO, it is difficult to keep the diversity of the population. The search rates are relatively low, and hence, it may require more computation time while solving the complex optimization problem. Like genetic algorithm, PSO algorithm also does not guarantee an optimal solution.

6.2.2 Society Civilization Algorithm

The concept of society civilization algorithm proposed by (Ray and Liew, 2003) is based on complex intra- and inter-society interactions which are derived out of two fundamental observations from nature:

    i. A civilization emerges and advances due to cooperative relationships among its societies.
    ii. Individuals in a society interact with each another with an aim to improve.

A society refers to a set of mutually interacting individuals, while a civilization is a collection of all such societies at any given point of time. Individuals are fundamental social entities that interact with each other in a quest to improve. The goal is to improve the performance of an individual in a society to improve the performance of the society, which in turn will improve the civilization. Every society has a society leader, and the improvement in an individual's performance is due to a meaningful information acquisition from the leader of the society to which the individual belongs (intra-society interaction). The performance of civilization is due to the civilization leader. The society leader of each society will then improve its performance through acquisition of meaningful information from the civilization leader (inter-society interaction). Such an inter-society information exchange among leaders results in the migration of leaders to developed societies led by better-performing leaders. This process evolves to convert the poor-performing societies into better-performing societies.

    The algorithm begins with randomly generated initial population which represents initial state of civilization. The societies are then formed by gathering like-minded individuals in the society. The collection of the societies represents a civilization. The combined objective function taking into consideration the objective function and all constraints is then evaluated for every individual, which basically represents the performance of the individual. Another feature of the algorithm is that the leaders of different societies compete among themselves to attract members from other societies in order to expand their social boundary.

    In case of particle swarm optimization, the performance of the swarm improved if the individuals were allowed to stereotype (i.e. align themselves towards their group's central tendencies). The social civilization algorithm exploits the benefits of a faster change in an individual's performance through inter-society and intra-society interactions.

6.2.3 Civilized Swarm Optimization

Civilized swarm optimization (Selvakumar and Thanushkodi, 2009) is an integration of particle swarm optimization (PSO) and society civilization algorithm (SCA). Due to inter-society and intra-society search, the SCA provides a good balance between local search and global search. Hence, the SCA has the capability to handle exploration and exploitation of the most promising areas of these searched areas efficiently in a well-balanced manner. The PSO algorithm has the capability to use the gathered information about the search space to reach to the global optima. CSO algorithm thus combines the benefits of SCA and PSO algorithm and hence is expected to offer fast

convergence and better accuracy of solution. Now, let us consider a simple example
to demonstrate the steps of civilized swarm optimization.

$$\text{Minimize } Z = 2x_1^2 x_2 - 3x_1 x_2^2; \text{such that } 0 \le x_1 \le 5 \text{ and } 2 \le x_2 \le 6 \qquad (6.3)$$

Step 1: Determine the algorithm-specific parameters:
   In CSO algorithm, the algorithm-specific parameters are the number of
particles (N), the number of societies (N_s), inertia weights (w), and accelera-
tion constants (c_1 and c_2). Let us set these parameters as: $N = 16$, $N_s = 4$,
$w = 0.60$, $c_1 = 1.60$, and $c_2 = 1.70$.
Step 2: Generate initial population:
   An initial population having 16 solutions is randomly generated, as
shown in Table 6.1.
Step 3: Arrange the individuals in ascending order of their fitness values
(Table 6.2).
Step 4: Determine the civilization leader and society leaders:
   Since the problem is of minimization, the solution No. 12 is the civiliza-
tion leader. As there are four societies, there will be four society leaders
represented by solution Nos. 12, 11, 3, and 2.
Step 5: Group the members to form societies:
   The members of civilization are then grouped to form societies (in
this particular case, four societies). The criterion to form the groups is the
Euclidean distance between the members. The members are grouped as
given in Table 6.3.

TABLE 6.1
Initial Population

Sol. No.	x_1	x_2	Z	v_1	v_2
1	3.90	2.17	100.64	1.42	2.87
2	1.19	2.00	23.69	1.73	4.33
3	0.09	4.51	18.33	1.37	3.47
4	4.05	3.22	136.49	4.54	2.56
5	2.40	4.54	49.02	1.55	4.15
6	3.01	2.90	73.62	2.32	3.07
7	4.30	2.66	137.11	1.11	3.13
8	4.29	4.19	176.36	0.40	2.01
9	3.22	2.12	71.50	0.42	3.16
10	3.02	4.35	81.66	3.02	3.62
11	0.73	3.45	16.83	3.84	4.42
12	0.17	4.99	16.30	4.46	2.83
13	3.22	4.61	95.04	3.56	2.34
14	2.33	3.86	48.11	0.02	3.17
15	2.51	4.71	53.44	4.89	3.00
16	1.84	4.62	27.54	0.59	3.74

TABLE 6.2

Arranging Individuals in Ascending Order of Their Fitness Values

Sol. No.	x_1	x_2	Z	v_1	v_2
12	0.17	4.99	16.30	1.42	2.87
11	0.73	3.45	16.83	1.73	4.33
3	0.09	4.51	18.33	1.37	3.47
2	1.19	2.00	23.69	4.54	2.56
16	1.84	4.62	27.54	1.55	4.15
14	2.33	3.86	48.11	2.32	3.07
5	2.40	4.54	49.02	1.11	3.13
15	2.51	4.71	53.44	0.40	2.01
9	3.22	2.12	71.50	0.42	3.16
6	3.01	2.90	73.62	3.02	3.62
10	3.02	4.35	81.66	3.84	4.42
13	3.22	4.61	95.04	4.46	2.83
1	3.90	2.17	100.64	3.56	2.34
4	4.05	3.22	136.49	0.02	3.17
7	4.30	2.66	137.11	4.89	3.00
8	4.29	4.19	176.36	0.59	3.74

TABLE 6.3

Grouping of Members to Form Societies

Civilization Leader (CL)	Society Number	Society Leaders (SL)	Society Members
Solution No. 12	1	Solution No. 12	Solution Nos. 12, 15, 16
	2	Solution No. 11	Solution Nos. 8, 10, 11, 13, 14
	3	Solution No. 3	Solution Nos. 3, 5
	4	Solution No. 2	Solution Nos. 1, 2, 4, 6, 7, 9,

Step 6: Update the civilization leader:

The civilization leader is represented by the global best solution in the entire population. The civilization also improves its performance so as to convert the poor-performing civilization into a better-performing civilization. Equations (6.4) and (6.5) are used to update the velocity (V_{CL}) and position (X_{CL}), respectively, of a civilization leader.

$$V_{CL\_i+1} = w \times V_{CL\_i} + c_1 \times r_1 \times \left(pBest_{CL\_i} - X_{CL\_i} \right) \tag{6.4}$$

$$X_{CL\_i+1} = X_{CL\_i+1} + V_{CL\_i+1} \tag{6.5}$$

Table 6.4 shows the updated position of the civilization leader obtained by using Eqs. (6.4) and (6.5).

TABLE 6.4

Updated Position of Civilization Leader

Sol. No.	x_1	x_2	Z
12	1.02	5.38	7.21

Step 7: Update the society leaders:

Society leaders (*SL*) also improve their performance through interaction with the civilization leader (*CL*) and also through memorizing their own best position (*p*Best$_{SL}$) attained till that stage. Equations (6.6) and (6.7) are used to update the velocity (*V*$_{SL}$) and position (*X*$_{SL}$), respectively, of a society leader.

$$V_{SL\_i+1} = w \times V_{SL\_i} + c_1 \times r_1 \times \left(pBest_{SL\_i} - X_{SL\_i}\right) + c_2 \times r_2 \times \left(X_{CL\_i} - X_{SL\_i}\right) \quad (6.6)$$

$$X_{SL\_i+1} = X_{SL\_i+1} + V_{SL\_i+1} \quad (6.7)$$

Table 6.5 shows the updated position of society leaders obtained by using Eqs. (6.6) and (6.7).

Step 8: Update the society members:

Society members (*SM*) are then updated through a useful interaction with the society leader (*SL*) and also based on their own best position achieved (*p*Best$_{SM}$). Equations (6.8) and (6.9) are used to update the velocity (*V*$_{SM}$) and position, respectively, of a society member (*X*$_{SM}$). *w* is the inertia coefficient, and c_1 and c_2 are the acceleration constants as mentioned in PSO algorithm. r_1 and r_2 are the random numbers generated between 0 and 1.

$$V_{SM\_i+1} = w \times V_{SM\_i} + c_1 \times r_1 \times \left(pBest_{SM\_i} - X_{SM\_i}\right) + c_2 \times r_2 \times \left(X_{SM\_i} - X_{SL\_i}\right) \quad (6.8)$$

$$X_{SM\_i+1} = X_{SM\_i} + V_{SM\_i+1} \quad (6.9)$$

Table 6.6 shows the updated position of society members obtained by using Eqs. (6.8) and (6.9).

TABLE 6.5

Updated Position of Society Leaders

Sol. No.	x_1	x_2	Z
12	1.02	5.38	7.21
11	1.88	5.87	17.50
3	1.05	5.38	7.37
2	3.80	4.40	136.75

TABLE 6.6
Updated Position of Society Members

Sol. No.	x_1	x_2	Z
16	2.77	6.00	58.08
14	3.69	4.92	131.71
5	2.89	6.00	66.58
15	1.96	6.00	18.59
9	3.95	4.66	152.22
6	4.99	4.26	247.08
10	4.07	6.00	171.85
13	1.66	6.00	9.74
1	5.00	5.73	285.64
4	3.63	5.38	127.62
7	4.07	6.00	172.08
8	2.72	6.00	55.16

It can be seen that after the first iteration, the best solution has a function value (Z) = 7.21, which is better than that of the initial population (i.e. Z = 16.30)

Step 9: Repeat Steps 2–8 till convergence.

6.2.4 OPPOSITION-BASED LEARNING CIVILIZED SWARM OPTIMIZATION

Since no information is available to choose initial solutions, usually the decision-makers start with random guesses. Obviously, the computation time is directly related to the distance of the guess from the optimal solution. Thus, by probability, out of the total number of solutions, 50% of the solutions are farther to the optimum solution. The opposite of such solutions will naturally be closer to the optimum solution. Thus, we find the opposite solution of every solution in the population and the better of these two solutions will be considered for subsequent steps. Since in CSO algorithm, we compute velocities also randomly for initial population, we find the opposite of the velocities also. Out of the original and opposite velocities, the one which produces a higher improvement in the updated solution will be considered. The opposite solutions (and velocities) are calculated for all generations till convergence. Thus, with opposition-based learning (Tizhoosh, 2005), it is expected that the algorithm will converge fast and will provide better accuracy of solution. If L and U are the lower and upper bounds, respectively, then the opposite $(\tilde{x})$ of a solution (x) is obtained by using Eq. (6.10).

$$\tilde{x}_i = L_i + U_i - x_i \tag{6.10}$$

If V_{max} and V_{min} are the minimum and maximum velocities, then the opposite of a velocity is obtained by using Eq. (6.11)

$$\tilde{v}_i = V_{i\_max} + V_{i\_min} - v_i \tag{6.11}$$

TABLE 6.7

Opposite Solutions

Sol. No.	$\tilde{x}_1$	$\tilde{x}_2$	z	$\tilde{v}_1$	$\tilde{v}_2$
1	1.1	5.83	3.78	3.58	5.13
2	3.81	6	144.13	3.27	3.67
3	4.91	3.49	212.61	3.63	4.53
4	0.95	4.78	11.24	0.46	5.44
5	2.6	3.46	59.04	3.45	3.85
6	1.99	5.1	28.83	2.68	4.93
7	0.7	5.34	7.89	3.89	4.87
8	0.71	3.81	15.46	4.6	5.99
9	1.78	5.88	14.35	4.58	4.84
10	1.98	3.65	36.55	1.98	4.38
11	4.27	4.55	180.48	1.16	3.58
12	4.83	3.01	186.90	0.54	5.17
13	1.78	3.39	31.77	1.44	5.66
14	2.67	4.14	62.78	4.98	4.83
15	2.49	3.29	54.24	0.11	5
16	3.16	3.38	85.15	4.41	4.26

Now, let us evaluate the effect of opposition-based learning on the solution for the same example discussed in Section 6.3.4. Table 6.7 shows the opposite of the original solutions $(\tilde{x}_i)$ and the opposite of the velocities $(\tilde{v}_i)$ obtained by using Eqs. (6.10) and (6.11).

The better out of the original and opposite solutions (marked bold) is then selected, as shown in Table 6.8.

It can be seen from Table 6.8 that out of 16 solutions, eight solutions are selected from the original solutions and the opposite solutions each. The average fitness value of the original population is 70.33, whereas that for the population of opposite solutions is 25.19 only. This clearly shows that with opposition-based learning, the solutions are much better than the randomly generated population, thus having more potential to explore the search space and fast convergence. In a similar manner, the opposite of initially generated velocities will also be evaluated and those which provide more improvement in the objective function will be considered. It should be noted that the opposite of the solutions and velocities are to be obtained not only for the initial population but also for every subsequent generation. However, the effect of opposition-based learning can be diminishing over the generations. Now, to demonstrate the effectiveness of the opposition-based learning CSO, two application examples from the manufacturing field are discussed in the subsequent sections.

6.3   APPLICATION EXAMPLES

Now, to demonstrate the effectiveness of opposition-based learning CSO algorithm, two case studies related to parametric optimization of two different manufacturing processes are considered.

TABLE 6.8

Initial Population Consisting of Better Solution Out of the Original and Opposite Solutions

Sol. No.	Original Solution			Opposite Solution			Finally Selected Solution		
	x_1	x_2	Z	$\tilde{x}_1$	$\tilde{x}_2$	Z	x_1	x_2	Z
1	3.90	2.17	100.64	1.1	5.83	3.78	1.1	5.83	3.78
2	1.19	2.00	23.69	3.81	6	144.13	1.19	2	23.69
3	0.09	4.51	18.33	4.91	3.49	212.61	0.09	4.51	18.28
4	4.05	3.22	136.49	0.95	4.78	11.24	0.95	4.78	11.24
5	2.40	4.54	49.02	2.6	3.46	59.04	2.4	4.54	49.02
6	3.01	2.90	73.62	1.99	5.1	28.83	1.99	5.1	28.83
7	4.30	2.66	137.11	0.7	5.34	7.89	0.7	5.34	7.89
8	4.29	4.19	176.36	0.71	3.81	15.46	0.71	3.81	15.46
9	3.22	2.12	71.50	1.78	5.88	14.35	1.78	5.88	14.35
10	3.02	4.35	81.66	1.98	3.65	36.55	1.98	3.65	36.55
11	0.73	3.45	16.83	4.27	4.55	180.48	0.73	3.45	16.83
12	0.17	4.99	16.30	4.83	3.01	186.90	0.17	4.99	16.30
13	3.22	4.61	95.04	1.78	3.39	31.77	1.78	3.39	31.77
14	2.33	3.86	48.11	2.67	4.14	62.78	2.33	3.86	48.11
15	2.51	4.71	53.44	2.49	3.29	54.24	2.51	4.71	53.44
16	1.84	4.62	27.54	3.16	3.38	85.15	1.84	4.62	27.54

6.3.1 OPTIMIZATION OF ABRASIVE WATER JET MACHINING (AWJM) PROCESS

The AWJM process uses a high-velocity water jet in combination with abrasive particles for cutting different types of materials using a set-up as shown in Figure 6.1.This process thus combines the benefits of both abrasive jet machining and water jet machining. The main advantages of this process are minimal thermal damage, omnidirectional cutting, cutting without delamination, and the ability to cut complex profiles. However, for cutting hard materials with the AWJM, the power requirement is very high, which puts a limit on the material removal rate increasing the time required to cut a part.

Tuning the process parameters appropriately so as to improve the material removal rate keeping in view the constraint on power consumption is one of the ways to deal with this issue. An optimization model for abrasive water jet machining formulated based on the analysis given by Hashish (1989) and Jain et al. (2007) is presented below.

6.3.2 OBJECTIVE FUNCTION

The objective is to maximize the material removal rate (Z_1) as given by Eq. (12).

$$\text{Maximize } Z_1 = d_{\text{awn}} f_n (h_c + h_d) \tag{6.12}$$

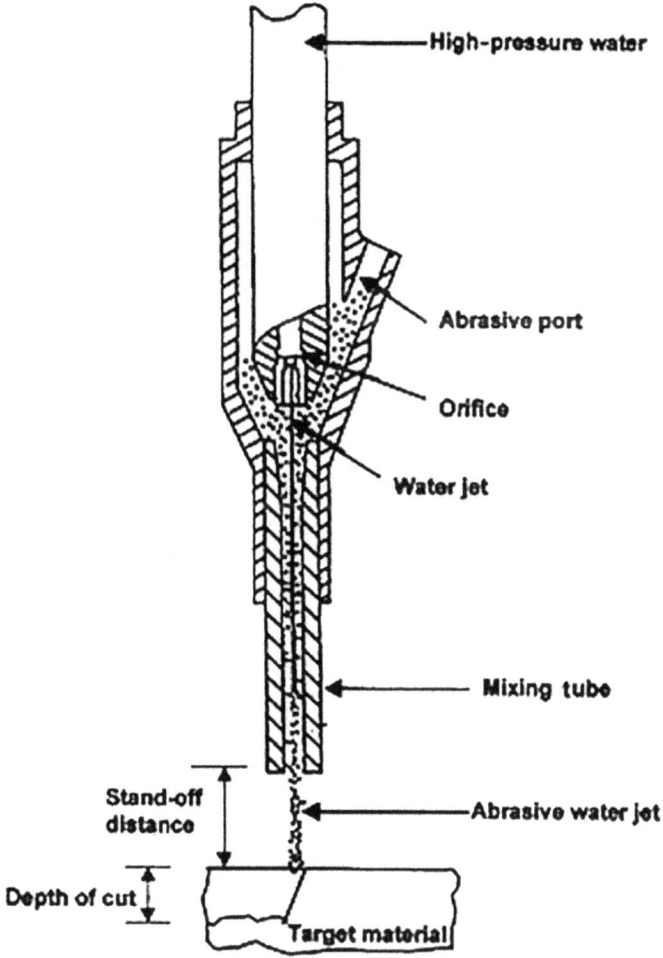

FIGURE 6.1 General scheme of abrasive water jet machining (AWJM) process (Jain and Jain, 2001).

where 'h_c' is the indentation depth due to cutting wear and is given by Eq. (13).

$$h_c = \left(\frac{1.028 \times 10^{4.5}\varsigma}{C_k^{1/3} f_r^{0.4}} \right)\left(\frac{d_{awn}^{0.2} M_a^{0.4}}{f_n^{0.4}} \right)\left(\frac{M_w P_w^{0.5}}{M_a + M_w} \right) - \left(\frac{18.48 K_a^{2/3}\varsigma^{1/3}}{C_k^{1/3} f_r^{0.4}} \right)\left(\frac{M_w P_w^{0.5}}{M_a + M_w} \right)^{1/3} ;$$

if $\alpha_t \leq \alpha_0$.   (6.13)

$$h_c = 0, \text{ if } \alpha_t \geq \alpha_0.$$ (6.14)

'h_d' is the indentation depth due to deformation wear and is given by Eq. (15).

h_d

$$= \frac{\eta_a d_{awn} M_a \left[K_1 M_w P_w^{0.5} - (M_a + M_w) v_{ac} \right]^2}{\left(1570.8\sigma_{fw}\right) d_{awn}^2 f_n (M_a + M_w)^2 + \left(K_1 C_{fw} \eta_a\right) \left[K_1 M_w P_w^{0.5} - (M_a + M_w) v_{ac} \right] M_a M_w P_w^{0.5}}$$

(6.15)

$$\alpha_0 \approx \left(\frac{0.02164 C_K^{1/3} f_r^{0.4}}{K_a^{2/3} \xi^{1/3}} \right) \left(\frac{\dot{M}_a + \dot{M}_w}{\dot{M}_w P_w^{0.5}} \right)^{1/3} \text{(degrees)}.$$

(6.16)

$$\alpha_t \approx \left(\frac{0.389 \times 10^{-4.5} \rho_a^{0.4} C_K}{\xi} \right) \left(\frac{d_{awn}^{0.8} f_n^{0.4} \left(\dot{M}_a + \dot{M}_w \right)}{\dot{M}_a^{0.4} \dot{M}_w P_w^{0.5}} \right) \text{(degrees)}.$$

(6.17)

$$v_{ac} = 5\pi^2 \frac{\sigma_{cw}^{2.5}}{\rho_a^{0.5}} \left[\frac{1 - v_a^2}{E_{Ya}} + \frac{1 - v_w^2}{E_{Yw}} \right]^2 \text{(mm/s)}$$

(6.18)

$$K_1 = \sqrt{2} \times 10^{4.5} \xi.$$

(6.19)

$$C_K = \sqrt{3000 \sigma_{fw} f_r^{0.6} / \rho_a} \text{ (mm/s)}$$

(6.20)

$$Ka = 3.$$

6.3.2.1 Constraint

The constraint is on power consumption as given by Eq. (6.21).

$$Z_2 = 1.0 - \frac{P_w \cdot M_w}{P_{\max}} \geq 0$$

(6.21)

A combined objective function (Z) is formulated considering both the objective function (Z_1) and the constraint function (Z_2) as given by Eq. (6.22).

$$\text{Min. } Z = -Z_1 - \text{Penalty} \times Z_2$$

(6.22)

The description of various symbols that appear in Eqs. (6.12)–(6.22) is provided in Table 6.9.

TABLE 6.9

Description of Notations (Jain et al., 2007)

Notation	Description	Unit	Value
ρ_a	Density of abrasive particles	kg/mm³	3.95×10^{-6}
ν_a	Poisson ratio of abrasive particles		0.25
E_{Ya}	Modulus of elasticity of abrasive particles	MPa	350,000
f_r	Roundness factor of abrasive particles		0.35
f_s	Sphericity factor of abrasive particles		0.78
η_a	Proportion of abrasive grains effectively participating in machining		0.70
ν_w	Poisson ratio of work material		0.20
E_{Yw}	Modulus of elasticity of work material	MPa	114,000
σ_{ew}	Elastic limit of work material	MPa	883
σ_{fw}	Flow stress of work material	MPa	8142
C_{fw}	Drag friction coefficient of work material		0.002
ξ	Mixing efficiency between abrasive and water		0.8
P_{max}	Allowable power consumption value	kW	56

6.3.2.2 Variable Bounds

The five decision variables considered for this model are water jet pressure at the nozzle exit (P_w), diameter of abrasive water jet nozzle (d_{awn}), feed rate of nozzle (f_n), mass flow rate of water (M_w), and mass flow rate of abrasives (M_a). The bounds for the five variables are given in Eqs. (6.23)–(6.27).

$$50 \le P_w \le 400 \,(\text{MPa}) \tag{6.23}$$

$$0.5 \le d_{awn} \le 5 \,(\text{mm}) \tag{6.24}$$

$$0.2 \le f_n \le 25 \,(\text{mm/s}) \tag{6.25}$$

$$0.02 \le M_w \le 0.2 \,(\text{Kg/s}) \tag{6.26}$$

$$0.0003 \le M_a \le 0.08 \,(\text{Kg/s}) \tag{6.27}$$

Now, the steps of opposition-based learning CSO discussed in Sections 6.3.4 and 6.3.5 are implemented to solve the above optimization model.

Step 1: Determine the algorithm-specific parameters:
    Let us set these parameters as: $N = 15$, $N_s = 4$, $w = 0.65$, $c_1 = 1.65$, and $c_2 = 1.75$.

Step 2: Generate initial population:
   An initial population having 15 solutions (chosen from the best among the randomly generated solutions and their opposite solutions) is chosen as shown in Table 6.10.

Step 3: Arrange the individuals in ascending order of their fitness values:
   Table 6.11 shows the initial population arranged in ascending order of the fitness values.

Step 4: Determine the civilization leader and society leaders:
   Since the problem is of minimization, the solution No. 15 is the civilization leader. As there are four societies, there will be four society leaders represented by solutions Nos. 15, 7, 9, and 11.

Step 5: Group the members to form societies:
   The members of civilization are then grouped to form societies (in this particular case, four societies). The criterion to form the groups is the Euclidean distance between the members. The members are grouped as given in Table 6.12.

Step 6: Update the civilization leader:
   The civilization leader is represented by the global best solution in the entire population. The civilization also improves its performance so as to convert the poor-performing civilization into a better-performing civilization. Equations (6.4) and (6.5) are used to update the velocity (V_{CL}) and position (X_{CL}), respectively, of a civilization leader. The solutions are updated by considering the opposite solutions and opposite velocities.
   Table 6.13 shows the updated position of the civilization leader obtained by using Eqs. (6.4) and (6.5).

TABLE 6.10
Initial Population

Sol. No.	P_w	d_{awn}	f_n	M_w	M_a	z
1	136.03	1.90	11.62	0.42	0.015	−8.97
2	312.26	3.15	14.90	0.24	0.070	−0.31
3	206.18	4.49	9.04	0.05	0.032	−16.49
4	150.44	2.29	10.16	1.02	0.001	471.73
5	123.78	0.68	20.26	1.35	0.027	503.80
6	242.22	1.61	14.48	0.36	0.006	139.91
7	213.86	0.80	13.05	0.26	0.055	−72.89
8	89.32	1.27	1.19	1.53	0.044	356.13
9	131.69	1.49	14.24	0.26	0.055	−63.28
10	156.86	4.22	8.27	0.89	0.068	340.55
11	194.38	3.06	17.66	0.14	0.056	−38.35
12	161.17	3.20	13.93	0.64	0.056	171.23
13	204.05	3.93	15.39	0.07	0.046	−24.98
14	80.74	4.48	21.53	0.70	0.023	−9.54
15	285.97	1.97	14.98	0.23	0.070	−109.71

TABLE 6.11

Arranging Individuals in Ascending Order of Their Fitness Values

Sol. No.	P_w	d_{awn}	f_n	M_w	M_a	Z
15	285.97	1.97	14.98	0.23	0.070	−109.71
7	213.86	0.80	13.05	0.26	0.055	−72.89
9	131.69	1.49	14.24	0.26	0.055	−63.28
11	194.38	3.06	17.66	0.14	0.056	−38.35
13	204.05	3.93	15.39	0.07	0.046	−24.98
3	206.18	4.49	9.04	0.05	0.032	−16.49
14	80.74	4.48	21.53	0.70	0.023	−9.54
1	136.03	1.90	11.62	0.42	0.015	−8.97
2	312.26	3.15	14.90	0.24	0.070	−0.31
6	242.22	1.61	14.48	0.36	0.006	139.91
12	161.17	3.20	13.93	0.64	0.056	171.23
10	156.86	4.22	8.27	0.89	0.068	340.55
8	89.32	1.27	1.19	1.53	0.044	356.13
4	150.44	2.29	10.16	1.02	0.001	471.73
5	123.78	0.68	20.26	1.35	0.027	503.8

TABLE 6.12

Grouping of Members to Form Societies

Civilization Leader (CL)	Society Number	Society Leaders (SL)	Society Members
Solution No. 12	1	Solution No. 15	Solution Nos. 2, 15
	2	Solution No. 7	Solution Nos. 6, 7
	3	Solution No. 9	Solution Nos. 1, 4, 5, 8, 9
	4	Solution No. 11	Solution Nos. 3, 10, 11, 12, 13, 14

TABLE 6.13

Updated Position of Civilization Leader

Sol. No.	P_w	d_{awn}	f_n	M_w	M_a	Z
15	307.72	2.23	17.50	0.28	0.071	−57.06

Step 7: Update the society leaders:

Society leaders (*SL*) also improve their performance through an interaction with the civilization leader (*CL*) and also through memorizing their own best position (*p*Best$_{SL}$) attained till that stage. Equations (6.6) and (6.7) are used to update the velocity (V_{SL}) and position (X_{SL}), respectively, of a society leader.

Table 6.14 shows the updated position of society leaders obtained by using Eqs. (6.6) and (6.7).

TABLE 6.14

Updated Position of Society Leaders

Sol. No.	P_w	d_{awn}	f_n	M_w	M_a	Z
15	307.72	2.23	17.50	0.28	0.071	−57.06
7	198.27	1.32	15.36	0.22	0.068	−89.76
9	168.97	1.64	24.12	0.32	0.059	−100.45
11	230.56	2.57	22.04	0.19	0.063	−56.94

Step 8: Update the society members:

Society members (*SM*) are then updated through a useful interaction with the society leader (*SL*) and also based on their own best position achieved (*p*Best$_{SM}$). Equations (6.8) and (6.9) are used to update the velocity (*V$_{SM}$*) and position, respectively, of a society member (*X$_{SM}$*).

Table 6.15 shows the updated position of society members obtained by using Eqs. (6.8) and (6.9).

Since the objective here is to minimize the combined objective function, it can be seen from Table 6.15 that after the first iteration, the best solution is solution No. 5 having a function value (*Z*) = −143.11, which is better than that of the initial population (i.e. *Z* = −109.71). Thus, for the next generation, solution No. 5 is the civilization leader and solution Nos. 5, 3, 9, and 7 are the society leaders for the four societies, respectively.

Step 9: Repeat Steps 2 to 8 till convergence.

TABLE 6.15

Updated Position of Society Members

Sol. No.	P_w	d_{awn}	f_n	M_w	M_a	Z
13	121.62	1.14	5.29	0.79	0.035	154.18
3	315.02	1.10	14.79	0.18	0.080	−118.43
14	150.73	4.56	2.54	0.06	0.049	−14.70
1	51.35	2.10	15.61	1.10	0.070	−21.05
2	256.65	0.14	1.89	1.89	0.004	−13.68
6	102.36	0.50	12.99	0.62	0.080	−16.67
12	133.15	1.73	6.60	0.01	0.027	−1.83
10	111.52	0.50	25.00	0.07	0.034	−13.14
8	96.21	1.40	18.05	0.01	0.059	−0.83
4	96.90	2.42	12.16	0.41	0.062	−31.62
5	216.66	3.99	5.70	0.24	0.051	−143.11

6.3.3 RESULTS OF OPTIMIZATION OF AWJM PROCESS USING OPPOSITION-BASED CSO ALGORITHM

The optimum solution is obtained as the best value of ten independent runs. At convergence, the optimum solution obtained by using opposition-based learning CSO is shown in Table 6.16.

Table 6.17 shows the comparative performance of the opposition-based learning CSO with that reported in the literature by using genetic algorithm and particle swarm optimization algorithm.

It can be seen from Table 6.17 that the opposition-based learning CSO algorithm outperformed genetic algorithm and particle swarm optimization, showing a significant improvement in the objective function of about 65% over GA and 10% over PSO.

6.3.4 OPTIMIZATION OF CNC TURNING PROCESS

This example presents a model considered by Kim et al. (2007) which was used in multi-pass turning operation of a mild steel workpiece using a carbide tool. The objective of this model is to minimize the production cost in dollars/piece. The objective function is given below:

$$\text{Min. Cost} = n \times \left(\frac{3141.59}{V.f.d} + \frac{2.879 \times 10^{-8} \times V^4 \times f^{0.75}}{d^{0.025}} + 10 \right) \qquad (6.28)$$

In the above equation, 'n' is the number of passes and 'd' is the depth of cut. The values of $n = 2$ and $d = 2.5$ are considered as the same are used in the previous studies. The allowable ranges for speed 'V' and feed rate 'f' are given as:

$$50 \leq V \leq 400\,\text{m/min} \qquad (6.29)$$

$$0.30 \leq f \leq 0.75\,\text{mm/rev} \qquad (6.30)$$

TABLE 6.16
Optimum Solution Obtained by Using Opposition-Based Learning CSO

Method	P_w	d_{awn}	f_n	M_w	M_a	Z
Opposition-based learning CSO	395.01	4.89	6.529	0.1418	0.08	−256.18

TABLE 6.17
Optimum Solution Obtained by Using Opposition-Based Learning CSO

Method	P_w	d_{awn}	f_n	M_w	M_a	Z
GA (Jain et al., 2007)	398.3	3.726	23.17	0.141	0.079	−90.257
PSO	400	3.242	13.084	0.142	0.080	−230.50
Opposition-based learning CSO	395.01	4.89	6.529	0.1418	0.08	−256.18

There are four physical constraints in this model as mentioned below.

a. *Cutting force constraint:*

It is necessary to put a constraint on the cutting force to limit the deflection of the workpiece or the cutting tool, which would result in dimensional errors, and to reduce the power required for the cutting process. The cutting force constraint is given in terms of maximum force '$F_{c\max}$'.

$$\text{Cutting force}(F_c) \le F_{c\max} \tag{6.31}$$

where

$$F_c = \left(28.10 \times V^{0.07} - 0.525 \times V^{0.5}\right) \cdot d \times f \left\{ 1.59 + 0.946 \left\{ \frac{1+x}{\left[(1-x)^2 + x \right]^{0.5}} \right\} \right\} \tag{6.32}$$

$F_{c\max} = 85\,\text{kg}.$

$$x = \left[\frac{V}{142} \times e^{2.21 \times f} \right]^2 \tag{6.33}$$

b. *Power constraint:*

The power required during the cutting operation (P_c) should not exceed the available power ($P_{c\max}$) of the machine tool. Thus, the cutting power constraint is given by:

$$P_c \le P_{c\max} \tag{6.34}$$

where

$$P_c = \frac{0.746 \times F_c \times V}{4500} \tag{6.35}$$

and $P_{c\max} = 2.25\,\text{kW}.$

c. *Minimum and maximum tool-life limits:*

A lower tool life limit may be imposed by the production system either because of the tool supply or because of the limit of number of tools in the tool magazine. Similarly, feasible feed and speed domain put limits on the higher tool life. The limit for the tool life (*TL*) is as given below:

$$25 \le TL \le 45. \tag{6.36}$$

d. *Tool–chip interface temperature constraint:*

The constraint on permissible tool–chip interface temperature (*T*) is given as:

$$T \le 1000°C$$

where

$$T = 132 \times V^{0.4} \times f^{0.2} \times d^{0.105} \qquad (6.37)$$

Now, the steps of the opposition-based learning CSO are implemented to solve the above optimization problem. The following algorithm-specific parameters are considered.
- Maximum number of iterations: 150.
- Inertia weight factor (w): 0.60.
- Acceleration coefficients: $c_1 = 1.85$ and $c_2 = 1.65$.

6.3.5 RESULTS OF OPTIMIZATION OF CNC TURNING PROCESS USING OPPOSITION-BASED CSO ALGORITHM

Table 6.18 shows the optimum cutting parameter data using CSO along with the published results using genetic algorithm. From the results shown in Table 6.18, it is observed that for multi-pass turning operation, the cost per piece obtained using the opposition-based CSO algorithm is 79.536 $/piece. Table 6.18 shows that the opposition-based CSO algorithm provides better accuracy of result than genetic algorithm.

6.4 CONCLUSIONS

In this chapter, the concept of opposition-based learning civilized swarm optimization was applied to two different problems related to machining. In both cases, it was observed that the opposition-based learning CSO provides better accuracy of solution than other popular algorithms such as genetic algorithm and particle swarm optimization. This is mainly due to the fact that the CSO algorithm combines the benefits of PSO and SCA and opposition-based learning improves our chance to start with a closer (fitter) solution by checking the opposite solution simultaneously. The implementation of opposition-based learning CSO algorithm reveals the fact that it is a very robust algorithm having a very efficient searching ability in the multi-minima environment with a relatively high convergence rate as it provides clustered search that results in better exploration and exploitation of the search space. The algorithm can also be easily modified to suit the optimization of process parameters of other machining processes.

However, as civilized swarm optimization integrates PSO and SCA, the number of control parameters required to be tuned is more than that of both PSO and SCA. Hence, several experimentations are required to determine the optimum values of the control parameters of CSO. Secondly, the incorporation of opposition-based learning increases the number of function evaluations in every generation and thus requires more computational efforts.

TABLE 6.18
Optimal Machining Conditions for CNC Turning Process

Method	V	f	Cost ($/pc)
GA (Kim et al., 2007)	147.71	0.3614	79.569
Opposition-based learning CSO	148.22	0.3618	79.536

REFERENCES

Bergh F., Engelbrecht A. P. (2006) A study of particle swarm optimization particle trajectories. *Information Sciences*, Vol. 176, pp. 937–971.

Hashish M. (1989) A model for abrasive-waterjet (AWJ) machining. *Journal of Engineering Materials and Technology*, Vol. 111(2), pp. 154–162.

Holland J. H. (1975) *Adaptation in Neural and Artificial Systems*. University of Michigan Press, Ann Arbour, MI.

Jain N.K., Jain V.K. (2001) Modeling of material removal in mechanical type advanced machining processes: A state of the art review. *International Journal of Machine Tools and Manufacture*, Vol. 41, pp. 1573–1635.

Jain N. K., Jain V. K., Deb K. (2007) Optimization of process parameters of mechanical type advanced machining processes using genetic algorithm. *International Journal of Machine Tools and Manufacture*, Vol. 47, pp. 900–919.

Kennedy J., Eberhart R. (1995) Particle swarm optimization. *Proceedings of IEEE International Conference on Neural Networks*, Vol. 4, pp. 1942–1948.

Kim S. S., Kim I., Mani V., Kim H. J. (2008) Real-coded genetic algorithm for machining condition optimization. *International Journal of Advanced Manufacturing Technology*, Vol. 38, pp. 894–895.

Ray T., Liew K. M. (2003) Society and civilization: An optimization algorithm based on the simulation of social behavior. *IEEE Transactions on Evolutionary Computation*, Vol. 7 (4), pp. 389–396.

Selvakumar I., Thanushkodi K. (2009) Optimization using civilized swarm: Solution to economic dispatch with multiple minima. *Electric Power Systems Research*, Vol. 79(1), pp. 8–16.

Tizhoosh H. R. (2005) Opposition-Based Learning: A New Scheme for Machine Intelligence, *International Conference on Computational Intelligence for Modeling Control and Automation-CIMCA'2005*, Vienna, Austria, pp. 695–701.

7 Application of Particle Swarm Optimization Method to Availability Optimization of Thermal Power Plants

Hanumant P. Jagtap
Department of Mechanical Engineering,
Zeal College of Engineering and Research

Anand K. Bewoor
Cummins College of Engineering for Women

Firozkhan Pathan
D. Y. Patil Institute of Technology

Ravinder Kumar
Lovely Professional University

CONTENTS

7.1 INTRODUCTION

Continuous supply of electricity is the need of the society in developing countries such as India. The industrialization sector is completely reliant on a stable power supply. The thermal power plant (TPP) is one of the major sources of electricity generation. Therefore, the TPP should be highly reliable and able to produce electricity with maximum capacity. However, the sudden failure of any major critical equipment of the TPP may result in huge power generation loss [1]. Earlier researchers have focused their study on the reliability and availability investigation of critical equipment of the system used in different applications [2–5].

Kuo et al. [6] examined the optimal availability in the steady state under three different repairable systems that were exposed to a server failure. Based on the obtained results, the manager can determine the best stationary availability in their repairable system with their existing data to our numerical model. Therefore, achieving a high level of availability is often essential. The numerical analysis can, therefore, be used to identify situations in which the repair time distributions and system parameters achieve a better cost–benefit ratio.

James et al. [7] analyzed the growth of the system reliability by using the actual field failure data. The main objective of the enhancements in system reliability is to improve the performance during the demonstration to achieve the predicted or contractually required commitment for system reliability. An effective reliability growth model can be used to predict when the reliability goal can be achieved based on the previous reliability performance.

Wenbin et al. [8] studied the maintenance strategy of corrective maintenance in combination with the overhaul. It was assumed that the errors that have occurred before the set overhaul time be remedied by corrective maintenance and repair be the effect of corrective maintenance and minimal or incomplete overhaul. Therefore, the Markov reward approach was implemented to determine the total cost associated with the calculation of reliability of a machine tool caused by operation and maintenance activities.

Inoue and Yamada [9] investigated the approach to modeling software reliability based on the effects of the change point, and the incomplete debugging environment by the growth process of software reliability was described with a semi-Markov process. In addition, some useful measures to evaluate the software reliability and parameter estimation of our model were discussed.

Okafor et al. [10] carried out the availability assessment of steam and gas turbine units of a thermal power plant using the Markovian approach. The availability analysis of the system was performed to determine the effect of failure and repair rates of each subsystem on the overall performance of that system. The data for the failure rate and repair rate were calculated from the results of the analysis of the operating data; a maximum loss in all the units was considered to be planned.

With the objective of maximizing the availability of the TPP with continuous power generation, it is necessary to investigate the availability of selected equipment/subsystem/system of the TPP. In such cases, two important research areas can be identified to conduct the further study as follows:

1. Identification and ranking of critical equipment of thermal power plants.
2. Selection of appropriate maintenance strategy for the selected critical equipment of the thermal power plant.

This study presents the use of Markov-based probabilistic approach for the availability analysis of the thermal power plant. The maintenance data or failure data are collected from the maintenance history of the thermal power plant. In this study, the whole thermal power plant is divided into six major subsystems, viz. boiler air circulation system, coal supply subsystem, boiler furnace subsystem, turbine generator subsystem, condenser subsystem, and water circulation subsystem. The critical equipment of the selected subsystems of the TPP is used for the availability analysis, viz. primary air fan (PA), boiler drum, stacker reclaimer, condenser, turbine-governing system, and boiler feedwater pump (BFP). The results obtained from the availability simulation model are used for ranking the critical equipment and for recommending the suitable maintenance strategy for the thermal power plant. Further, the particle swarm optimization method is adopted for optimizing the availability parameters of the TPP. The detailed system description is discussed next.

7.2   SYSTEM DESCRIPTION

The selected pieces of equipment of the six subsystems of TPP for the development of Markov-based availability simulation model are (i) primary air fan "A," (ii) boiler drum "B," (iii) stacker reclaimer "C," (iv) condenser "D," (v) turbine-governing system "E," and (vi) boiler feedwater pump "F."

The primary air fan and boiler feedwater pump consist of two units running in parallel combinations. Therefore, the failure of any one of these units would result in running the system at reduced capacity. However, the other pieces of equipment such as boiler drum, stacker reclaimer, condenser, and turbine-governing system consist of a single unit. The failure of any one of these units would result in complete failure of the TPP.

7.2.1   ASSUMPTIONS

a. Failure and repair rates for each subsystem are constant and statistically independent.
b. Not more than one failure occurs at a time.
c. A repaired unit is as good as new, performance-wise.
d. The standby units are of the same nature and capacity as the active units.

7.2.2   NOMENCLATURE

◯ : Good capacity state.   ⬭ : Reduced capacity state.

□ : Failed state.

A, B, C, D, E, F: Equipment is in good operating state.
a, b, c, d, e: Indicates the failed state of *A, B, C, D, E*.
$\bar{A}$: Indicates the reduced capacity state of *A*.
$\bar{F}$: Indicates the reduced capacity state of *F*.
λi: Mean constant failure rate.
μ_i: Mean constant repair rate.
Pi(t): Probability that at time "t" the system is in the *i*th state.
': Derivative with respect to "*t*."

7.2.3 AVAILABILITY SIMULATION MODELING OF THERMAL POWER PLANTS

The proposed availability simulation model for a selected subsystem of the TPP is developed on the basis of Markov probabilistic approach. The new mathematical expressions using the Laplace transform technique are derived. The transition diagram presenting the different states of the system of the TPP is presented in Figure 7.1. It contains three different working states of the system, viz. working at full capacity, reduced capacity, and failed state. In transition diagram, a total of 20 states are presented, which include the states from "0" to "19." Of the states, the state "0" represents the system is working at full capacity, then the states from "1" to "3" designate the system working with reduced capacity, and the remaining states from

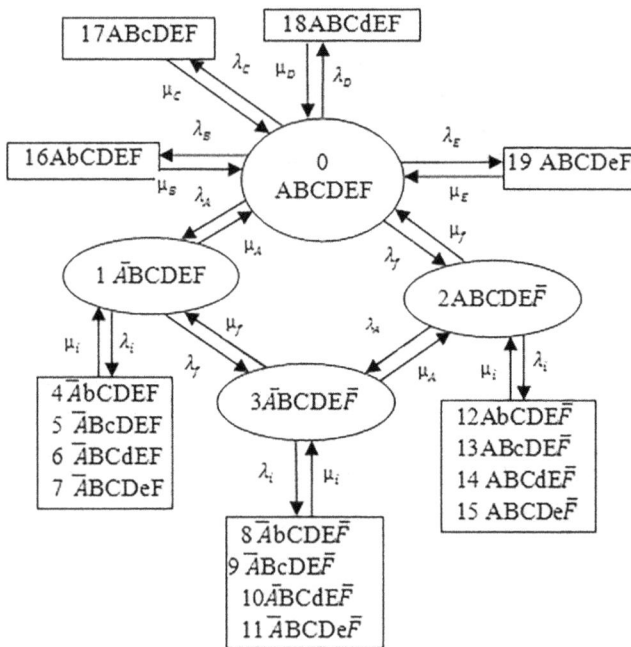

FIGURE 7.1 Transition diagram of a subsystem of a thermal power plant.

"4" to "19" represent the failed state of the system. The probability-based differential equations are derived using the Laplace transformation technique associated with the transition diagram.

$$P_0'(t) + (\lambda_A + \lambda_B + \lambda_C + \lambda_D + \lambda_E + \lambda_F) P_0(t)$$
$$= \mu_A P_1(t) + \mu_B P_{16}(t) + \mu_C P_{17}(t) + \mu_D P_{18}(t) + \mu_E P_{19}(t) + \mu_F P_2(t) \qquad (7.1)$$

$$P_1'(t) + (\lambda_B + \lambda_C + \lambda_D + \lambda_E + \lambda_F + \mu_A) P_1(t)$$
$$= \mu_B P_4(t) + \mu_C P_5(t) + \mu_D P_6(t) + \mu_E P_7(t) + \mu_F P_3(t) + \lambda_A P_0(t) \qquad (7.2)$$

$$P_2'(t) + (\lambda_A + \lambda_B + \lambda_C + \lambda_D + \lambda_E + \mu_F) P_2(t)$$
$$= \mu_A P_3(t) + \mu_B P_{12}(t) + \mu_C P_{13}(t) + \mu_D P_{14}(t) + \mu_E P_{15}(t) + \lambda_F P_0(t) \qquad (7.3)$$

$$P_3'(t) + (\lambda_B + \lambda_C + \lambda_D + \lambda_E + \mu_A + \mu_F) P_3(t)$$
$$= \mu_B P_8(t) + \mu_C P_9(t) + \mu_D P_{10}(t) + \mu_E P_{11}(t) + \lambda_A P_2(t) + \lambda_F P_1(t) \qquad (7.4)$$

$$P_4'(t) + \mu_B P_4(t) = \lambda_B P_1(t) \qquad (7.5)$$

$$P_5'(t) + \mu_C P_5(t) = \lambda_C P_1(t) \qquad (7.6)$$

$$P_6'(t) + \mu_D P_6(t) = \lambda_D P_1(t) \qquad (7.7)$$

$$P_7'(t) + \mu_E P_7(t) = \lambda_E P_1(t) \qquad (7.8)$$

$$P_8'(t) + \mu_B P_8(t) = \lambda_B P_3(t) \qquad (7.9)$$

$$P_9'(t) + \mu_C P_9(t) = \lambda_C P_3(t) \qquad (7.10)$$

$$P_{10}'(t) + \mu_D P_{10}(t) = \lambda_D P_3(t) \qquad (7.11)$$

$$P_{11}'(t) + \mu_E P_{11}(t) = \lambda_E P_3(t) \qquad (7.12)$$

$$P_{12}'(t) + \mu_B P_{12}(t) = \lambda_B P_2(t) \qquad (7.13)$$

$$P_{13}'(t) + \mu_C P_{13}(t) = \lambda_C P_2(t) \qquad (7.14)$$

$$P_{14}'(t) + \mu_D P_{14}(t) = \lambda_D P_2(t) \tag{7.15}$$

$$P_{15}'(t) + \mu_E P_{15}(t) = \lambda_E P_2(t) \tag{7.16}$$

$$P_{16}'(t) + \mu_B P_{16}(t) = \lambda_B P_0(t) \tag{7.17}$$

$$P_{17}'(t) + \mu_C P_{17}(t) = \lambda_C P_0(t) \tag{7.18}$$

$$P_{18}'(t) + \mu_D P_{18}(t) = \lambda_D P_0(t) \tag{7.19}$$

$$P_{19}'(t) + \mu_E P_{19}(t) = \lambda_E P_0(t) \tag{7.20}$$

With the use of initial conditions, at time $t = 0$, $P_i(t) = 1$ for $i = 0$; otherwise, $P_i(t) = 0$. Further, for long-run availability steady state, $\dfrac{d}{dt} \to 0$ and $t \to \infty$. The above equations can be modified as shown below.

$$(\lambda_A + \lambda_B + \lambda_C + \lambda_D + \lambda_E + \lambda_F)P_0 = \mu_A P_1 + \mu_B P_{16} + \mu_C P_{17} + \mu_D P_{18} + \mu_E P_{19} + \mu_F P_2 \tag{7.21}$$

$$(\lambda_B + \lambda_C + \lambda_D + \lambda_E + \lambda_F + \mu_A)P_1 = \mu_B P_4 + \mu_C P_5 + \mu_D P_6 + \mu_E P_7 + \mu_F P_3 + \lambda_A P_0 \tag{7.22}$$

$$(\lambda_A + \lambda_B + \lambda_C + \lambda_D + \lambda_E + \mu_F)P_2 = \mu_A P_3 + \mu_B P_{12} + \mu_C P_{13} + \mu_D P_{14} + \mu_E P_{15} + \lambda_F P_0 \tag{7.23}$$

$$(\lambda_B + \lambda_C + \lambda_D + \lambda_E + \mu_A + \mu_F)P_3 = \mu_B P_8 + \mu_C P_9 + \mu_D P_{10} + \mu_E P_{11} + \lambda_A P_2 + \lambda_F P_1 \tag{7.24}$$

$$\mu_B P_4 = \lambda_B P_1 \tag{7.25}$$

$$\mu_C P_5 = \lambda_C P_1 \tag{7.26}$$

$$\mu_D P_6 = \lambda_D P_1 \tag{7.27}$$

$$\mu_E P_7 = \lambda_E P_1 \tag{7.28}$$

$$\mu_B P_8 = \lambda_B P_3 \tag{7.29}$$

$$\mu_C P_9 = \lambda_C P_3 \tag{7.30}$$

$$\mu_D P_{10} = \lambda_D P_3 \tag{7.31}$$

$$\mu_E P_{11} = \lambda_E P_3 \tag{7.32}$$

$$\mu_B P_{12} = \lambda_B P_2 \tag{7.33}$$

$$\mu_C P_{13} = \lambda_C P_2 \tag{7.34}$$

$$\mu_D P_{14} = \lambda_D P_2 \tag{7.35}$$

$$\mu_E P_{15} = \lambda_E P_2 \tag{7.36}$$

$$\mu_B P_{16} = \lambda_B P_0 \tag{7.37}$$

$$\mu_C P_{17} = \lambda_C P_0 \tag{7.38}$$

$$\mu_D P_{18} = \lambda_D P_0 \tag{7.39}$$

$$\mu_E P_{19} = \lambda_E P_0 \tag{7.40}$$

Now, solving Eqs. (7.21) and (7.22) using the matrix method, we have

$P_1 = L_1 P_0$, $P_2 = L_2 P_0$, $P_3 = L_3 P_0$, $P_4 = K_B L_1 P_1$, $P_5 = K_C L_1 P_0$, $P_6 = K_D L_1 P_0$, $P_7 = K_E L_1 P_0$,

$P_8 = K_B L_3 P_0$, $P_9 = K_C L_3 P_0$, $P_{10} = K_D L_3 P_0$, $P_{11} = K_E L_3 P_0$, $P_{12} = K_B L_2 P_0$, $P_{13} = K_C L_2 P_0$,

$P_{14} = K_D L_2 P_0$, $P_{15} = K_E L_2 P_0$, $P_{16} = K_B P_0$, $P_{17} = K_C P_0$, $P_{18} = K_D P_0$, $P_{19} = K_E P_0$,

where

$$K_A = \frac{\lambda_A}{\mu_A}, K_B = \frac{\lambda_B}{\mu_B}, K_C = \frac{\lambda_C}{\mu_C}, K_D = \frac{\lambda_D}{\mu_D}$$

The above equations are solved using the matrix method. Then, $\sum_{i=0}^{19} P_i = 1$.

$$P_0 = \left[\begin{array}{l} 1 + \sum_{i=1}^{3} L_i + K_B L_1 + K_C L_1 + K_D L_1 + K_E L_1 + K_B L_3 + K_C L_3 + K_D L_3 \\ \\ + K_E L_3 + K_B L_2 + K_C L_2 + K_D L_2 + K_E L_2 + K_B + K_C + K_D + K_E \end{array} \right]^{-1}$$

Now, the simulation model for the steady-state availability of TPP can be obtained as the summation of all the working state probabilities, i.e.,

$$A_V = \left[1 + L_1 + L_2 + L_3 \right] P_0 \tag{7.41}$$

7.3   RESULTS AND DISCUSSION OF MARKOV-BASED ANALYSIS

The availability of the TPP is majorly influenced by the failure rate and repair rate of its subsystems. Therefore, the failure rate and repair rate of the selected equipment were collected and analyzed for availability. On the basis of the Markov probabilistic approach, the performance of the TPP system was evaluated using the proposed

Markov availability simulation model. The various state probabilities are provided in the transition diagram, which assisted in developing the availability matrix for the selected equipment of the TPP. The availability of the equipment of the TPP is obtained and tabulated in Tables 7.1–7.6. Moreover, Figures 7.2–7.7 represent the effect of the failure rate and repair rate of the equipment on the overall availability of the TPP.

TABLE 7.1
Availability Matrix for PA of Unit 2

$\mu 1$	$\lambda 1$				
	7.6024E-05	8.00E-05	0.000105367	1.30708E-04	0.00015605
0.008962	0.9997	0.9997	0.9996	0.9996	0.9995
0.009434	0.9997	0.9997	0.9997	0.9996	0.9995
0.012421	0.9997	0.9997	0.9997	0.9997	0.9996
0.015409	0.9998	0.9998	0.9997	0.9997	0.9997
0.018396	0.9998	0.9998	0.9998	0.9997	0.9997

TABLE 7.2
Availability Matrix for Boiler Drum of Unit 2

$\mu 2$	$\lambda 2$				
	5.6954E-05	6.00E-05	7.89368E-05	9.79216E-05	0.00011691
0.004029	0.9682	0.9675	0.9632	0.9588	0.9545
0.004241	0.9689	0.9682	0.9641	0.9599	0.9558
0.005584	0.9719	0.9714	0.9682	0.9651	0.9619
0.006927	0.9738	0.9734	0.9708	0.9682	0.9657
0.00827	0.9751	0.9747	0.9726	0.9704	0.9682

TABLE 7.3
Availability Matrix for Stacker Reclaimer of Unit 2

$\mu 2$	$\lambda 2$				
	7.7497E-05	8.16E-05	0.000107409	1.33241E-04	0.00015907
0.008962	0.9912	0.9907	0.9879	0.9851	0.9823
0.009434	0.9916	0.9912	0.9885	0.9858	0.9832
0.012421	0.9935	0.9932	0.9912	0.9892	0.9871
0.015409	0.9947	0.9945	0.9928	0.9912	0.9895
0.018396	0.9956	0.9953	0.9939	0.9926	0.9912

TABLE 7.4

Availability Matrix for Condenser of Unit 2

			$\lambda 4$		
$\mu 4$	5.6954E-05	6.00E-05	7.89368E-05	9.79216E-05	0.00011691
0.003827	0.9733	0.9726	0.9679	0.9633	0.9587
0.004029	0.974	0.9733	0.9689	0.9645	0.9601
0.005304	0.9773	0.9767	0.9733	0.97	0.9666
0.00658	0.9793	0.9788	0.9761	0.9733	0.9706
0.007856	0.9806	0.9803	0.9779	0.9756	0.9733

TABLE 7.5

Availability Matrix for Turbine-Governing System of Unit 2

			$\lambda 5$		
$\mu 5$	5.6954E-05	6.00E-05	7.89368E-05	9.79216E-05	0.00011691
0.001515	0.9593	0.9575	0.9463	0.9353	0.9246
0.001595	0.961	0.9593	0.9486	0.9382	0.9279
0.0021	0.9689	0.9676	0.9594	0.9512	0.9432
0.002605	0.9739	0.9728	0.966	0.9593	0.9527
0.00311	0.9772	0.9763	0.9706	0.9649	0.9593

TABLE 7.6

Availability Matrix for BFP of Unit 2

			$\lambda 6$		
$\mu 6$	8.1473E-05	8.58E-05	0.000112919	1.40076E-04	0.00016723
0.011393	0.9992	0.9991	0.9991	0.999	0.999
0.011993	0.9992	0.9992	0.9991	0.999	0.999
0.01579	0.9992	0.9992	0.9992	0.9991	0.9991
0.019588	0.9992	0.9992	0.9992	0.9992	0.9991
0.023386	0.9992	0.9992	0.9992	0.9992	0.9992

Figure 7.2 represents the effect of failure and repair rates of the PA on the overall system availability. It reveals that as the failure rate of PA increases from 7.6024E-05 (failures/h) to 0.00015605 (failures/h), the availability decreases by about 0.02%. Similarly, as the repair rate of PA increases from 0.008962 (repairs/h) to 0.018396 (repairs/h), the system availability increases by about 0.02%.

Figure 7.3 represents the effect of failure and repair rates of the boiler drum on the overall system availability. It reveals that as the failure rate of boiler drum increases

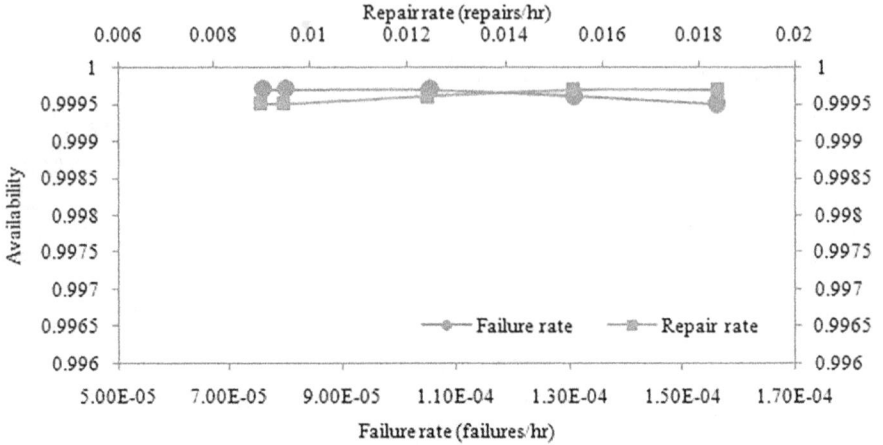

FIGURE 7.2 Effect of failure and repair rates of PA on system availability.

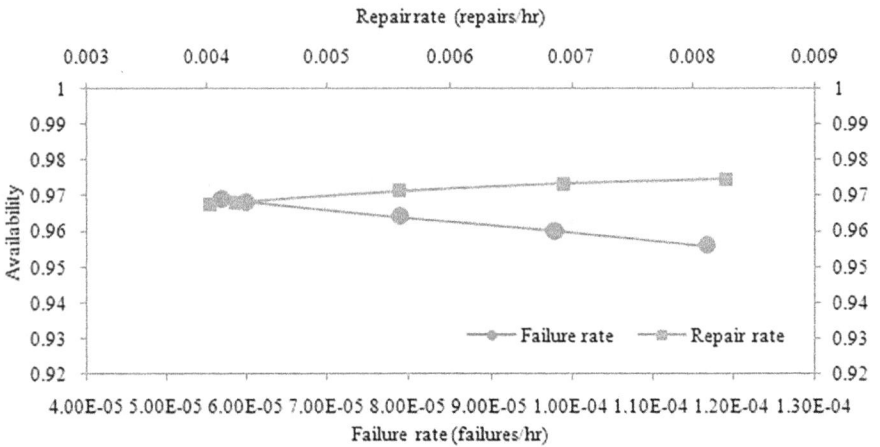

FIGURE 7.3 Effect of failure and repair rates of boiler drum on system availability.

from 5.6954E-05 (failures/h) to 0.00011691 (failures/h), the availability decreases by about 1.31%. Similarly, as the repair rate of boiler drum increases from 0.004029 (repairs/h) to 0.00827 (repairs/h), the system availability increases by about 0.72%.

Figure 7.4 represents the effect of failure and repair rates of the stacker reclaimer on the overall system availability. It reveals that as the failure rate of stacker reclaimer increases from 7.7497E-05 (failures/h) to 0.00015907(failures/h), the availability decreases by about 0.84%. Similarly, as the repair rate of stacker reclaimer increases from 0.008962 (repairs/h) to 0.018396 (repairs/h), the system availability increases by about 0.46%.

Figure 7.5 represents the effect of failure and repair rates of the condenser on the overall system availability. It reveals that as the failure rate of condenser increases

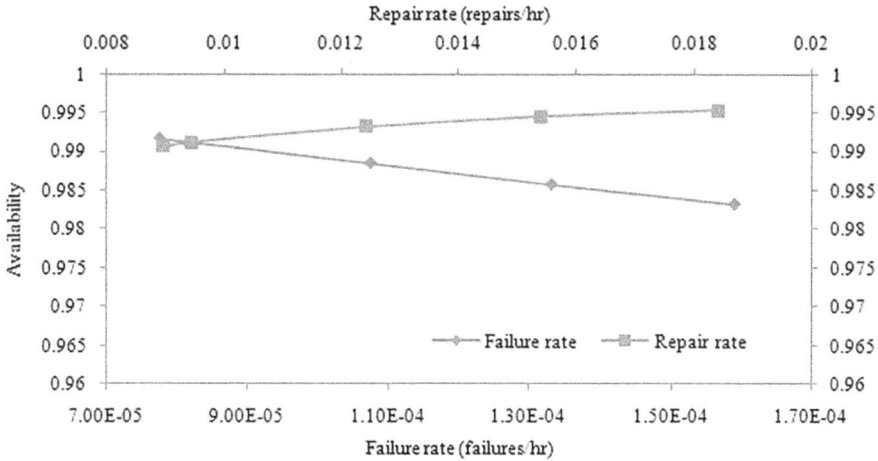

FIGURE 7.4   Effect of failure and repair rates of stacker reclaimer on system availability.

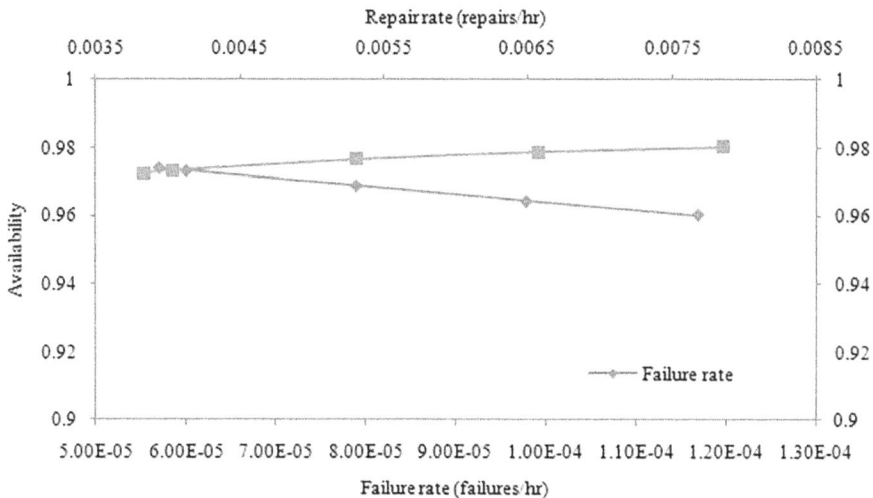

FIGURE 7.5   Effect of failure and repair rates of condenser on system availability.

from 5.6954E-05 (failures/h) to 0.00011691 (failures/h), the availability decreases by about 1.39%. Similarly, as the repair rate of condenser increases from 0.003827 (repairs/h) to 0.007856 (repairs/h), the system availability increases by about 0.77%.

Figure 7.6 represents the effect of failure and repair rates of the turbine-governing system on the overall system availability. It reveals that as the failure rate of the turbine-governing system increases from 5.6954E-05 (failures/h) to 0.00011691(failures/h), the availability decreases by about 3.31%. Similarly, as the repair rate of the turbine-governing system increases from 0.001515 (repairs/h) to 0.00311 (repairs/h), the system availability increases by about 1.88%.

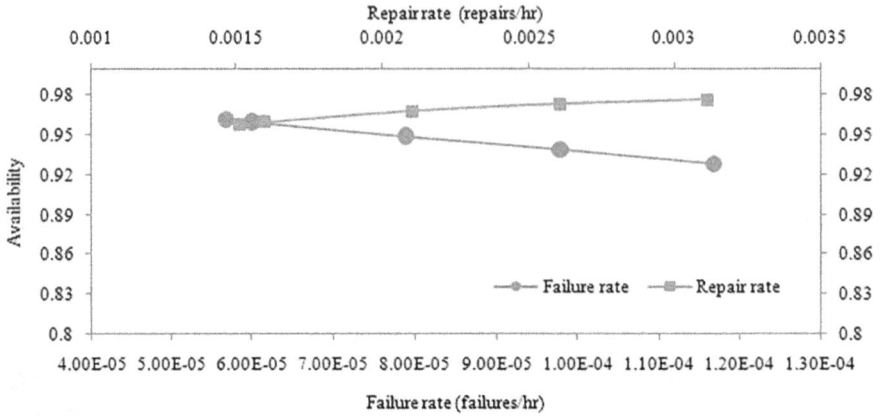

FIGURE 7.6 Effect of failure and repair rates of turbine-governing system on system availability.

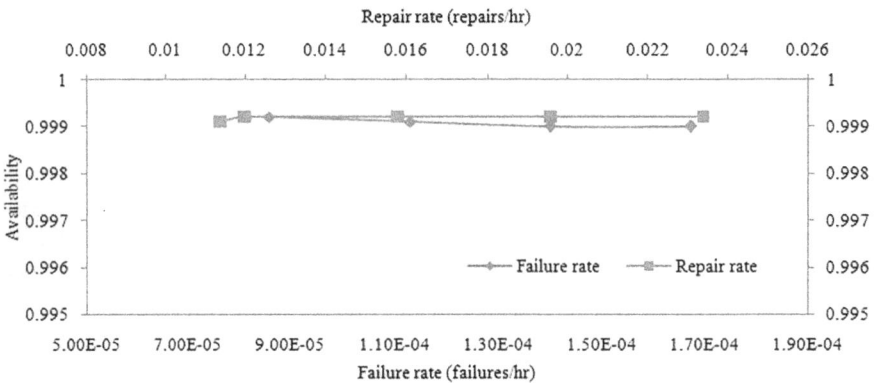

FIGURE 7.7 Effect of failure and repair rates of BFP on system availability.

Figure 7.7 represents the effect of failure and repair rates of the BFP on the overall system availability. It reveals that as the failure rate of BFP increases from 8.1473E-05(failures/h) to 0.00016723 (failures/h), the availability decreases by about 0.01%. Similarly, as the repair rate of BFP increases from 0.011393 (repairs/h) to 0.023386 (repairs/h), the system availability increases by about 0.02%.

The optimum availability is arrived at using the possible combination of failure rate and repair rate of the TPP so the plant will be available maximum as tabulated in Table 7.7.

It is observed from Table 7.7 that the failure of the turbine-governing system of the TPP affected the performance of the same rapidly and reduced the overall plant availability. Similarly, the failure of condenser and boiler drum affected the plant availability significantly.

TABLE 7.7
Optimum Values of Availability Parameters

Equipment	Failure Rate (λ_i)	Repair Rate (μ_i)	Decrease in Av due to λ_i (%)	Increase in Av due to μ_i (%)	Maximum Availability %
PA	8.00256E-05	0.009433962	0.02	0.02	99.97
Boiler drum	5.9952E-05	0.004241282	1.31	0.72	97.51
Stacker reclaimer	8.15763E-05	0.009433962	0.84	0.46	99.56
Condenser	5.9952E-05	0.004028648	1.39	0.77	98.06
Turbine-governing system	5.9952E-05	0.001594896	3.31	1.18	97.72
BFP	8.5761E-05	0.01199262	0.01	0.02	99.92

The availability parameters of such critical equipment of the TPP can be used for selecting the appropriate maintenance strategy using the optimization method. The next section describes the application of the PSO method to availability optimization.

7.4 PARTICLE SWARM OPTIMIZATION (PSO) TO OPTIMIZE THE AVAILABILITY OF TPPs

Complex technical problems can be solved by modern optimization methods. The modern optimization methods such as genetic algorithm, simulated annealing, ant colony optimization, particle swarm optimization, and neural network-based methods have been developed by previous researchers and adopted [11–13]. The applicability of such modern optimization techniques is rarely seen in the case of thermal power stations for the availability analysis in the published literature.

Jagtap et al. [14] developed a simulation modeling based on the availability of the boiler furnace system of a thermal power plant. The performance of the TPP system was evaluated based on Markov approach through a simulation model. The results obtained from the availability analysis and the corresponding optimized availability parameters such as failure rate and repair rate of the boiler furnace system were used for the availability optimization. In their study, the performance of the system was analyzed for a possible combination of failure rate and repair rate. Further, these parameters were used as inputs for optimizing the availability using particle swarm optimization method. The effect of particle number on the system availability was analyzed.

The study results showed that the maximum system availability could be achieved by 99.9845%. Moreover, the optimized values for the failure rate and repair rate of the subsystem were used to provide a suitable maintenance strategy for the boiler furnace system of the plant. The results of the study supported the decision-makers in the planning of maintenance activities according to the criticality of the subsystems for the allocation of resources.

Similarly, Jagtap et al. [15] performed the availability analysis of the coal supply system of a TPP using Markov probabilistic approach to birth and death of a thermal

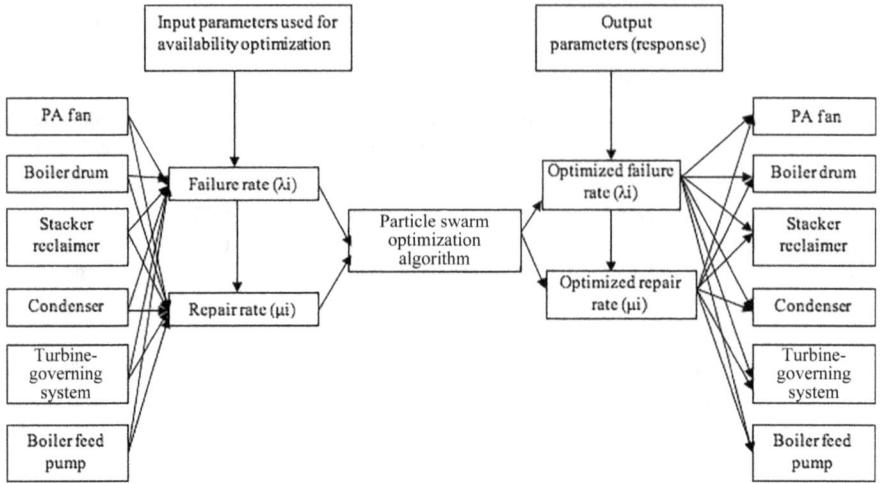

FIGURE 7.8   Methodology used for PSO-based availability optimization.

power plant. Study applies particle swarm algorithm to optimize parameters so as to enhance availability of the coal supply system, which in turn helps in the selection of appropriate maintenance strategy.

As less attention is paid in previous studies to the use of PSO method for availability optimization of the TPP, by reviewing some previous studies which focus on the development of Markov models for the performance evaluation and the use of failure parameters for availability optimization, an attempt is made in this study to make use of PSO method for the availability optimization of the selected subsystem of the TPP. The methodology used for PSO-based availability optimization is shown in Figure 7.8.

The objective of the PSO method is to optimize the availability of the TPP. The input parameters provided for the PSO algorithm are the failure rate and the repair rate of the selected equipment of the TPP. Then, using the best values Pbest and Gbest for particle position and velocity, the corresponding values of failure rate and repair rate can be obtained. The optimized values are further used for recommending suitable maintenance strategies for the equipment, such as condition-based maintenance, preventive maintenance, and breakdown maintenance.

7.5 CONCLUSION

In this study, a Markov-based availability simulation model was developed. The failure rate and repair rate of the subsystems of the TPP were taken as input. The availability matrices were developed, which facilities the maintenance person to decide on the maintenance priority according to the criticality level. The results obtained from the availability analysis revealed that the turbine-governing system, condenser, and boiler drum are the most critical equipment of the TPP from an availability point of view. On the basis of results obtained related to the optimum availability level mentioned in Table 7.7, the maintenance priority is recommended in the following order: (i) turbine-governing system, (ii) condenser, (iii) boiler drum, (iv) stacker reclaimer, (v) primary air fan, and (vi) boiler feedwater pump.

This study discussed the use of PSO method to find the optimized availability condition of the equipment used in the TPP. This study will help the maintenance person to identify the critical equipment of the TPP and to select the best maintenance strategy for the allocation of maintenance resources. From the optimized results obtained, it is concluded that the values obtained from PSO analysis are an important input for the selection of suitable maintenance strategy, which in turn improves the TPP availability.

REFERENCES

1. T. W. Huang, Y. Hsu, and W.F. Wu, Reliability analysis based on nonhomogeneous continuous-time markov modeling with application to repairable pumps of a power plant, *Int. J. Reliab. Qual. Saf. Eng.*, vol. 24, no. 1, pp. 1–14, 2017.
2. S. Du, Z. Zeng, L. Cui, and R. Kang, Reliability analysis of Markov history-dependent repairable systems with neglected failures, *Reliab. Eng. Syst. Saf.*, vol. 159, pp. 134–142, 2017.
3. J. Hassan, P. Thodi, and F. Khan, Availability analysis of a LNG processing plant using the Markov process, *J. Qual. Maint. Eng.*, vol. 22, no. 3, pp. 302–320, 2016.
4. S. P. Sharma, and Y. Vishwakarma, Application of Markov process in performance analysis of feeding system of sugar industry, *J. Ind. Math.*, vol. 2014, pp. 1–9, 2014.
5. L. Wang, Q. Yang, and Y. Tian, Reliability analysis of 6-component star Markov repairable system with spatial dependence, *Math. Probl. Eng.*, vol. 2017, no. 3, pp. 279–287, 2017.
6. C. Kuo, and J. Ke, Availability and comparison of spare systems with a repairable server, *Int. J. Reliab. Qual. Saf. Eng.*, vol. 26, no. 1, pp. 1–18, 2019.
7. L. James, G. Collins, and R. Govindarajulu, System reliability growth analysis during warranty, *Int. J. Math. Eng. Manag. Sci.*, vol. 4, no. 1, pp. 85–94, 2019.
8. Z. Wenbin, I. Frenkel, S. Guixiang, and I. Bolvashenkov, Markov reward approach and reliability associated cost model for machine tools maintenance-planning optimization, *Int. J. Math. Eng. Manag. Sci.*, vol. 4, no. 4, pp. 824–840, 2019.
9. S. Inoue, and S. Yamada, Markovian software reliability modeling with change-point, *Int. J. Reliab. Qual. Saf. Eng.*, vol. 25, no. 1, pp. 1–13, 2018.
10. C. E. Okafor, A. A. Atikpakpa, and U. C. Okonkwo, Availability assessment of steam and gas turbine units of a thermal power station using Markovian approach, *Arch. Curr. Res. Int.*, vol. 6, no. 4, pp. 1–17, 2016.
11. H. R. Maier, and A. R. Simpson, Power plant maintenance scheduling using Ant Colony Optimization – An improved formulation, *Eng. Optim.*, vol. 40, no. 4, pp. 309–329, 2008.
12. H. Garg, M. Rani, S. P. Sharma, and Y. Vishwakarma, Expert systems with applications intuitionistic fuzzy optimization technique for solving multi-objective reliability optimization problems in interval environment, *Expert Syst. Appl.*, vol. 41, no. 7, pp. 3157–3167, 2014.
13. C. Yao, B. Wang, and D. Chen, Reliability optimization of multi-state hydraulic system based on T-S fault tree and extended PSO algorithm, *IFAC Proc. Vol.*, vol. 46, no. 5, pp. 463–468, 2013.
14. H. Jagtap, A. Bewoor, R. Kumar, and M. Hossein, Markov-based performance evaluation and availability optimization of the boiler – furnace system in coal-fired thermal power plant using PSO, *Ener. Rep.*, vol. 6, pp. 1124–1134, 2020.

15. H. P. Jagtap, and A. K. Bewoor, Markov probabilistic approach-based availability simulation modeling and performance evaluation of coal supply system of thermal power plant. In: Varde, P., Prakash, R., Vinod, G. (eds.), *Reliability, Safety and Hazard Assessment for Risk-Based Technologies*. Lecture Notes in Mechanical Engineering, pp. 813–824. Springer, Singapore, 2020.

8 Optimization of Incremental Sheet Forming Process Using Artificial Intelligence-Based Techniques

Ajay Kumar and Deepak Kumar
Department of Mechanical Engineering,
Faculty of Engineering and Technology
Shree Guru Gobind Singh Tricentenary University

Parveen Kumar
Department of Mechanical Engineering,
Rawal Institute of Engineering and Technology

Vikas Dhawan
Shree Guru Gobind Singh Tricentenary University

CONTENTS

8.1 INTRODUCTION

In this modern era, the international market is accepting an alarming growth of obsolescence in the use of technology to manufacture components due to cutting-edge and customer-oriented requirements. Economical and customized products have become the basic requirement of manufacturing sector in order to upgrade their business [1]. Batch-type and customized productions require technologies that can produce components whose cycle time and set-up cost are significantly lower. Incremental sheet forming (ISF) has the capacity to produce sheet material components at a lower cost due to the flexibility and elimination of costly dies [2,3]. ISF evolved in the beginning of the current century and is still developing as a forming methodology that has proved its significance in manufacturing and design of customized sheet material components. ISF is also suitable to form nonsymmetrical geometries without using expensive dies for manufacturing complex components of sheet metal [4–6]. As an example, replacing the fuselage parts of old aircrafts is a major issue in the aerospace sectors due to the unavailability of forming dies and punches of these components. These kinds of problems can be overcome by processes such as ISF, which is a purely die-less forming process. It requires minimum energy in order to form the components. It also significantly reduces the tooling costs as compared to other conventional sheet material processes where expensive dies are used to produce any product [7–9]. Hence, multi-variety components in small batches can be manufactured at a low cost with ISF technology, which prevents the limitations of traditional sheet material forming processes.

In addition, ISF renders indispensable factors such as lightweight components, high productivity, flexibility in shapes and sizes of parts, and greater preciseness of process for manufacturing customized components economically [10,11]. ISF involves progressive plastic deformation of material, which is under consideration, by the stratagem action of forming tool that is regulated by a numerically controlled mechanism. Generally, a computer numerical control (CNC) milling machine, articulated industrial robot or purpose-built ISF machines are being used for accomplishing ISF process [12].

Single point incremental forming (SPIF), a category of ISF, exhibits the characteristics of a 'die-less process' by removing the need for dedicated dies and punches, which, in turn, results in making this process reliable and suitable for manufacturing customized parts economically due to the involvement of lower cycle time and cost of tools [13]. The schematic of SPIF, deforming the blank contour by contour by chasing the predetermined tool path, is illustrated in Figure 8.1. The applications of

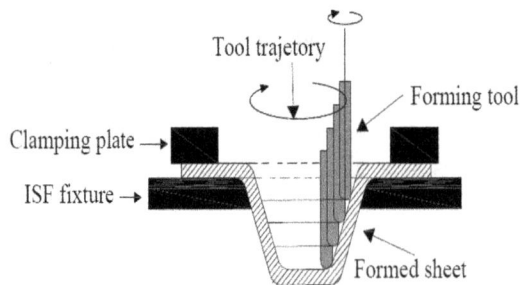

FIGURE 8.1 Single point incremental forming [1].

ISF process are very wide, including aerospace, automotive, architecture, medical and prototyping [14].

SPIF is beautified with the characteristics of the nature of local deformation, which enables this process to deform the sheet by a smaller amount of forming force as compared to conventional sheet forming methods. In addition, SPIF saves energy to a greater extent and reduces the set-up cost of process by employing the forming machinery of smaller size and capacity and decreasing the power consumption. Moreover, the ability of forming machinery, for employing SPIF technique, is delineated by the maximum forces that are needed for forming parts under particular input conditions of process and material. Hence, the investigation and prediction of forming force can ensure the effective and safe utilization of forming hardware [15].

In SPIF, the forming force can be measured by load cells or force dynamometers using various methods. Most of the researchers [14–23] have employed table-type dynamometers fixed between machine table and sheet clamping device or fixture. Kumar et al. [14] also studied the impact of various process variables on axial forming forces on AA2024 sheets using a forming tool of hemispherical shape. Forming forces were found to rise by 20.03% as tool radius increased from 3.76 to 7.83 mm for a wall angle of 64°. Kumar et al. [15] focused on the influence of interactions of tool radius–tool shape and wall angle–tool shape on forming forces on AA2024 sheets. An increase in tool radius (from 3.76 to 7.83 mm) resulted in a decrease in forces by 35.49%. Uheida et al. [16] studied the impact of feed rate and spindle speed on vertical forming forces on titanium grade 2 sheets of 0.8 mm thickness using a hemispherical end forming tool of 10 mm to produce varying wall angle conical frustum (VWACF). Alsamhan et al. [17] optimized the SPIF process for forming forces taking step size, feed rate, sheet thickness and tool radius into account on AA1050-H14 sheets. Furthermore, a model was established to estimate the forces by artificial intelligence techniques. Sakhtemanian et al. [18] developed a theoretical model for predicting the energy that can be converted to heat from the ultrasonic vibrations used in the study.

Long et al. [19] studied the influence of feed rate, material of sheet (AA1050-H14 and AA5052-H34), vibration amplitude (0, 6, 9, 12, 15 and 18 μm), ultrasonic power and tool radius on forming force and temperature induced during ultrasonic-assisted SPIF process. Honarpisheh et al. [20] investigated the forming force, thickness distribution and geometrical accuracy of bimetal sheets of AA1050 and copper (C-10100). The results showed that the axial forming forces increased from 1464 N and 1357 N to 1636 N and 1730 N for numerical and experimental tests, respectively, when the tool diameter was raised from 10 to 16 mm. Zhai et al. [21] investigated the effects of spindle speed, sheet thickness and step size on forming forces during ultrasonic-assisted ISF (U-ISF) and ISF on AA1050-O sheets to produce pyramidal frustums. Forming forces were found to reduce up to 40% when the ultrasonic amplitude was increased from 0 to 10 μm. Kumar and Gulati [22] studied the impact of various process variables on axial forming forces and optimized the SPIF process using Taguchi method for AA6063 and AA2024 sheets. The results showed that forming forces were found to be reduced by 14.20% when a hemispherical tool of 11.60 mm was used in place of a flat end tool of same diameter keeping other factors constant. Kumar and Gulati [23] studied the impact of various process variables on axial forming forces and thickness reduction and optimized the SPIF process using Taguchi method on AA2014 sheets. Conical frustums were formed using two different tool path approaches, viz. profile and helical, and these are well described by [1,13,22]. Chang et al. [24] developed an analytical model for predicting forming forces in single-pass SPIF, multi-pass SPIF and incremental hole flanging processes and validated the same with the experimental results by varying the impact factors such as sheet thickness, wall angle, step size and tool diameter for AA5052 and AA3003 alloy sheets.

It has also been observed in the literature [13,25,26] that the axial forming forces (Fz) are much greater than the other two components of forming forces (i.e. Fx and Fy). Therefore, secure implementation of forming machinery and forming tools can be ensured by determining the maximum axial forces required to execute the SPIF process. The knowledge about the effects of input factors on forming forces would lead to efficient prediction of failures of components as well as machinery used for executing the process. Furthermore, the prediction of forming forces helps in controlling the process online and estimating the power needed by forming machinery. The traditional models are not capable of describing the nonlinear relationship between the input parameters and the forming force. An artificial neural network (ANN) model can be used as a solution to formulate the complex and nonlinear relationship between the input and output factors [17]. The ANN model can also be applied for optimizing the performance and quality of response characteristics by choosing the appropriate input parameters. Furthermore, an ANN model can be trained using a data sample having inputs and the corresponding outputs [27]. In order to minimize the power consumption, control the process online and prevent the failure of hardware, the ANN-based model can be very useful for efficient prediction of forming force in SPIF.

Artificial intelligence-based models, using basic approaches, have been used in the literature of SPIF recently [28–31] and have not been proved accurate enough. A hybrid learning-based artificial neural network (HLANN) model can solve the issues encountered when using the basic ANN models and, hence, can be proposed to predict and optimize the maximum forming force developed during SPIF operation. In addition, two AI-based models, viz. a support vector machine (SVM)-based model and a Gaussian process regression (GPR)-based model, are also investigated. Both SVM and GPR models have been used by many researchers for prediction, and their results have been outstanding so far. These models have been very efficient, especially in situations where the training samples are very limited [32,33]. The results of these three models are then compared with the measured and experimental results to delineate the effectiveness of HLANN over the SVM and GPR. The purpose of this work is to investigate the impact of step size, wall angle and their interactions. To the best of authors' knowledge, the interactions of these parameters have not been studied on AA2024 sheets during SPIF process so far. Furthermore, the HLANN has also not been investigated during SPIF to predict the forming force. The prediction of forming force by using HLANN would open a new window for researchers and engineers to estimate the forming force and would make the investigation of the process economical by eliminating the cost involved in experimentation.

AA2024 is a popular aluminium alloy, which is known for its favourable characteristics in sheet metal applications, such as lightweight, high strength and

TABLE 8.1
Chemical Composition of AA2024 Sheet

Element	Al	Cr	Cu	Fe	Mg	Mn	Si	Ti	Zn
Weight %	91.50	0.10	4.60	0.30	1.70	0.80	0.50	0.10	0.20

TABLE 8.2
Levels of Input Variables Under Investigation

Variable	Level 1	Level 2	Level 3	Level 4
Wall angle (°)	60	64	68	-
Tool diameter(mm)	7.52	11.60	15.66	19.5

corrosion resistance. The chemical composition of this alloy is represented in Table 8.1. The procedure and instrument used for measuring these compositions are well explained in previous studies [2,15,22]. Table 8.2 depicts the varied impact factors and their levels. While varying the set of input factors, other parameters were kept constant according to a previous work [15].

8.2 MATERIALS AND METHODS

Figure 8.2 depicts the experimental set-up installed on CNC milling machine. The forming tool was firmly mounted on spindle to provide rotation to it, the SPIF fixture (which is hollow) was mounted on the machine table to clamp the sheet firmly and to provide a relative motion between the tool and the sheet, a load cell between the machine table and SPIF fixture was employed to measure the vertical downward force, and a data logger system was used to record the measured forces. The data logger system was assisted by Microscada software to store the real-time values of force. The working area of fixture was 200 mm × 200 mm. To produce conical frustums of 120 mm major diameter and 70 mm height, CAD model was designed in SolidWorks® software and the same was imported to Delcam™ software for generating numerical instructions for CNC milling machine, taking a helical tool path into consideration. Castrol Alpha SP 320 oil was used as a lubricant during forming operation.

8.2.1 DEVELOPMENT OF ANN MODEL TO PREDICT FORMING FORCE

AI-based models outperform the traditional methods, and ANN model is most useful in terms of its natural training process. The learning of ANN model and extraction of input features are executed automatically by adjusting the weights using back-propagation approach, while the process of feature extraction is dependent on statistical methods and done manually in all other AI/ML methods. In HLANN, a hybrid learning approach has been used to train ANN by using the input–output training

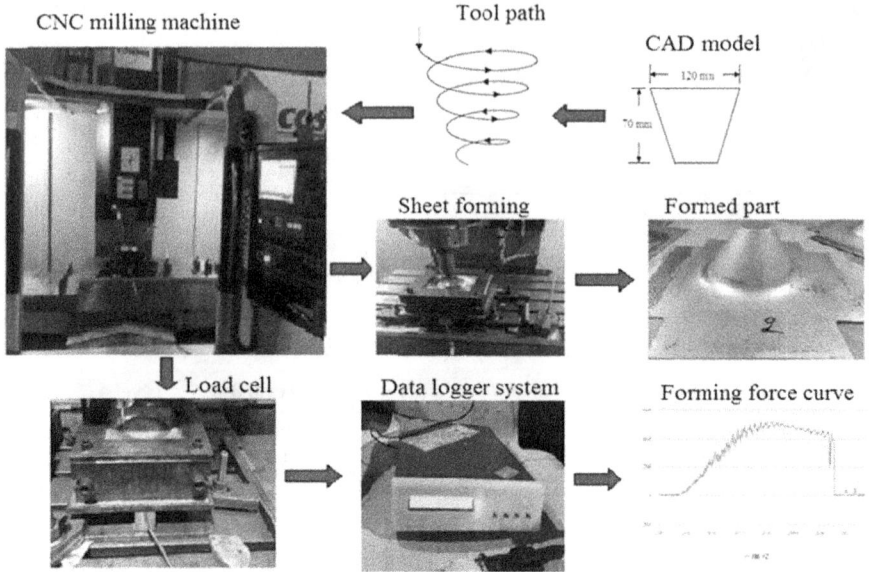

FIGURE 8.2   Experimental set-up and measurement of forming force.

data set based upon the experimental data. The ANN model consists of three layers, viz. input layer, hidden layer and output layer, as shown in Figure 8.3. The main purpose of this work is to explore the accuracy of the HLANN model by predicting the maximum forming force during SPIF process under given process conditions.

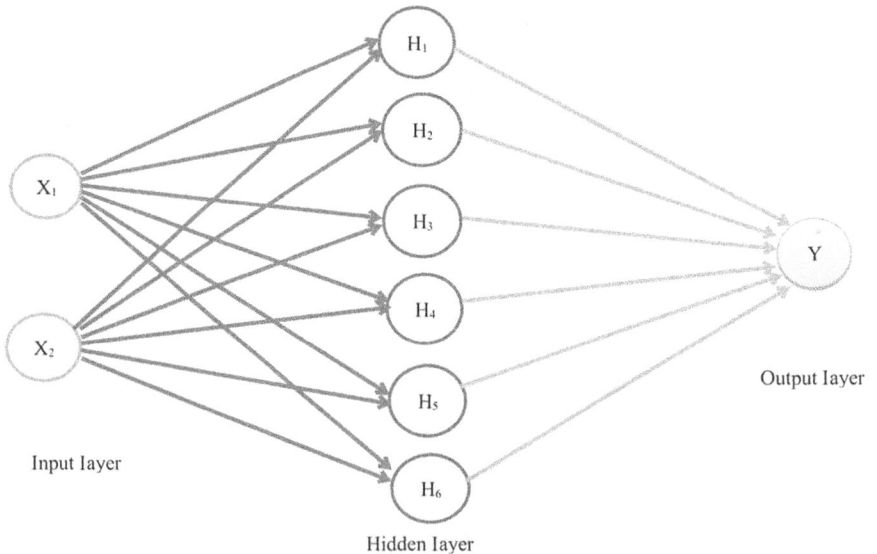

FIGURE 8.3   Artificial neural network model.

The ANN model, adopted in the current work, consists of an input layer composed of two neurons as inputs: x_1 and x_2. The weights are applied to each connection of neurons using a weight vector w, and the initial values of weights are chosen randomly. Besides these two inputs, one external input bias b is also applied. The charge v, induced at a neuron, is calculated by a summation function as follows:

$$v = \Sigma_i \, w_i x_i \text{ for all } i \tag{8.1}$$

An activation function $f()$ is then applied to the charge v to get the output y as follows:

$$y = f(v) \tag{8.2}$$

The output of input layers acts as input for the hidden layer, and this procedure continues till the output layer produces output Y for the first epoch. The output Y is compared with the expected output O, and an error value E is calculated as $E = O - Y$. The error E is propagated back to the network to modify the weights using backpropagation in order to minimize the error and fine-tune the ANN model. The change in output in accordance with the change in weights can be represented as derivation function expressed below:

$$\Delta y = \Sigma_i \frac{\partial y}{\partial wi} \Delta w_i + \frac{\partial y}{\partial b} \Delta b \text{ for all } i \tag{8.3}$$

The weights are changed using the gradient descent method, and a learning rate η is employed to vary the weights in each epoch till the error is minimized; the relation between the change in charge v, learning rate η and change in error is calculated by Eq. (8.4) as follows:

$$\Delta v = -\eta \Delta E \tag{8.4}$$

Similarly, the change in the weights, according to the error value, is calculated by Eq. (8.5) as follows:

$$\Delta w = w - \eta \, \frac{\partial E}{\partial w} \tag{8.5}$$

Furthermore, the new value of weight w_{new} for the next epoch is calculated by Eq. (8.6) as follows:

$$w_{new} = w + \Delta w \tag{8.6}$$

The training process of ANN model stops once the minimum error value is achieved, and the trained ANN model is then exposed to some new parameters to predict the output. The results of ANN model are considered to be reliable and of greater accuracy.

8.2.2 SUPPORT VECTOR MACHINE (SVM) MODEL

Support vector machine (SVM) is an AI algorithm that is trained by using the input–output pair as sample. SVM utilizes a maximum margin hyperplane linear model to give the maximum separation (w) between the distinguishable categories and hence maximizes a particular mathematical function with respect to a given data sample. The sample is represented in n-dimensional space (where n represents the number of features). The main elements of SVM are as follows: (i) maximum margin hyperplane, (ii) separating hyperplane, (iii) kernel function and (iv) soft margin. The maximum margin hyperplane provides the maximum separation distance amongst the data points so that the future data points can be predicted with a higher accuracy. The separating hyperplane is used to separate the data points with a decision boundary. Kernel function basically transforms the nonlinear feature space into a linear one. Soft margin is used to deal with wrongly separated data points, by the separating hyperplane. Figure 8.4 depicts the separation of data points with the maximum margin hyperplane and the separating hyperplane. SVM has capacity to outperform the basic AI-based models and yield a great accuracy [34]. Hence, an attempt has been made to compare the results of SVM model with the HLANN.

8.2.3 GAUSSIAN PROCESS REGRESSION (GPR) MODEL

GPR is a nonparametric and flexible method, based on AI, which finds out the probability distribution over all the functions fit in the given data in a linear regression model. GPR is very useful for exploring the high dimensionality of data and

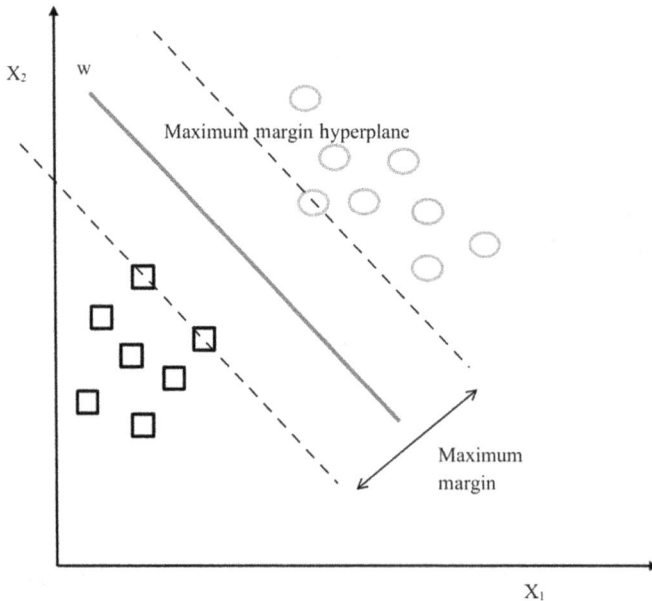

FIGURE 8.4   Support vector machine model.

multiple target values. For a training set $\{(x_i, y_i)$ for $i = 1,2, n)\}$ and a linear model $y = x^t + c$; where c is a constant, the kernel function for the GPR model can be given by Eq. (8.7) as follows:

$$K(x, x') = \sigma_f^2 \exp\left(-\tfrac{1}{2l}^2 \mod|x - x'|^2\right) \tag{8.7}$$

Here, σ^2 and l are the variance and length scale, respectively.

GPR has sought the attention of researchers because of its significance in measuring uncertainty of output data [35], especially when the data size is comparatively small. In the current work, the authors have employed GPR as the data set is of smaller size and compared the accuracy of GPR with the proposed model, i.e. HLANN.

8.3 RESULTS AND DISCUSSION

8.3.1 EXPERIMENTAL RESULTS AND ANALYSIS

The experimental measured values and predicted values of the maximal vertical downward forces (Fz_max.) are depicted in Table 8.3. The influence of interaction of wall angle and tool diameter is depicted by Figure 8.5. The maximal axial force increased dramatically when the amount of tool diameter and wall angle was increased because the larger wall angle results in deforming the material in large amount at an instant.

For a higher tool diameter, a large amount of material is deformed, which requires much forming load to produce local deformation. When a combination of higher levels of these two factors is employed, the forming force increases very rapidly and, hence, becomes the limiting factor of forming tool and machinery that should obviously be avoided. On the other hand, employment of a higher wall angle resulted

TABLE 8.3
Experimental and Estimated Results of Maximum Axial Force for the Given Conditions

Run	Tool Diameter	Wall Angle	Fz_max. (N) (Experimental)	Fz_max. (N) (HLANN)	Fz_max. (N) (SVM)	Fz_max. (N) (GPR)
1	7.52	60	764	765.00	808.17	784.13
2	7.52	64	806	808.77	863.29	836.70
3	7.52	68	889	884.99	934.62	904.77
4	11.6	60	907	910.04	964.29	927.49
5	11.6	64	1003	1005.12	1052.19	988.60
6	11.6	68	1133	1130.99	1180.31	1151.65
7	15.66	60	1114	1117.10	1136.86	1124.24
8	15.66	64	1251	1247.52	1281.30	1260.06
9	15.66	68	1434	1432.70	1459.96	1442.59
10	19.5	60	1369	1363.51	1319.82	1356.14
11	19.5	64	1553	1557.02	1512.18	1543.66
12	19.5	68	1784	1781.97	1733.76	1768.97

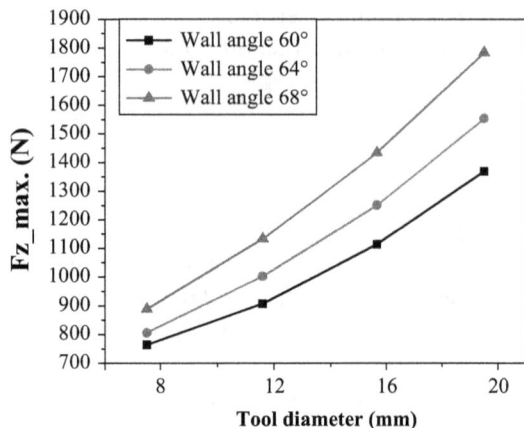

FIGURE 8.5 Influence of wall angle and tool diameter on maximal axial forces (experimental values).

in the fracture of sheet material well before achieving the designed height of conical frustums. Sheet material was fractured at a height of 55.6 mm when a combination of lower tool diameter (7.52 mm, in this case) and wall angle (68°, in this case) was employed. When the same wall angle (68°) was employed with a higher level of tool diameter (19.5 mm, in this case), sheet fractured at a relatively lower height (46.5 mm). This is due to the fact that excessive thinning of sheet occurred when the wall angle increases, and hence, less material is available to delay the fracture of thinner sheets. The combination of lower tool diameter and higher wall angle can deform the material by lower forming forces at the expense of formability.

8.3.2 PREDICTION OF AXIAL PEAK FORCES USING AI TECHNIQUES

The predicted results are observed for the proposed model (HLANN) and two state-of-the-art models, i.e. SVM and GPR. Both of the models (SVM and GPR) have been used by many researchers for prediction, and their results have been outstanding so far. These models have been very efficient, especially in situations where the training samples are of smaller size [31,32]. Hence, the results of these models are compared with the experimental results. Furthermore, the results predicted by HLANN are also compared. The prediction campaign is carried out using MATLAB 2018b on a system having i5–3320 CPU (64 bit), 8 GB memory.

8.3.3 HLANN USED FOR PREDICTION OF MAXIMUM AXIAL FORCE

The HLANN model, composed of three layers: an input layer, a hidden layer and an output layer, is shown in Figure 8.6. The input layer contains two neurons, i.e. one neuron for each of the input parameters. The hidden layer is composed of six neurons, and each neuron in this layer is connected to both the neurons in the input layer. Finally, the output layer contains one neuron to get a single output as the

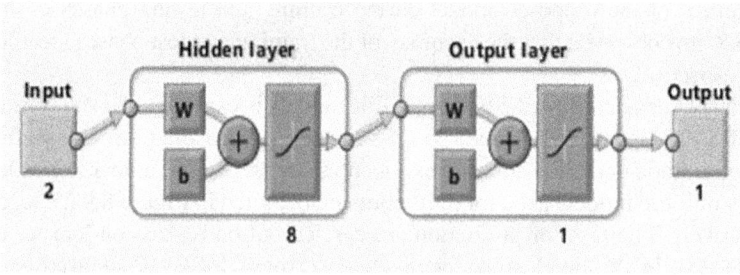

FIGURE 8.6 HLANN model used for predicting axial peak forces.

maximum axial force and all neurons of hidden layer are connected to the neuron in the output layer. The input vector to the model is comprised of weights, and the initial value is selected randomly. Based upon the multiplication of the strength of neuron and weights, the output is produced, which comprises of an activation function and a summation function. The output produced by one layer acts as the input for the subsequent layer, and finally, the desired output is obtained at the output layer. Based upon the difference between the observed and actual values, the error is calculated and weights in the hidden and input layers are altered to minimize the error and to tune the model.

The proposed model was trained on Levenberg–Marquardt backpropagation training algorithm. This algorithm requires less time to train the model, and the training stops automatically when the values of weights stop further improvement. The training statistics are depicted in Figure 8.7.

During the training phase, it is observed that the optimal value of the gradient was 0.80044 and the same was determined after 23 epochs. The learning rate was kept 0.001 per epoch, and a total of ten validation checks were executed as depicted in Figure 8.7.

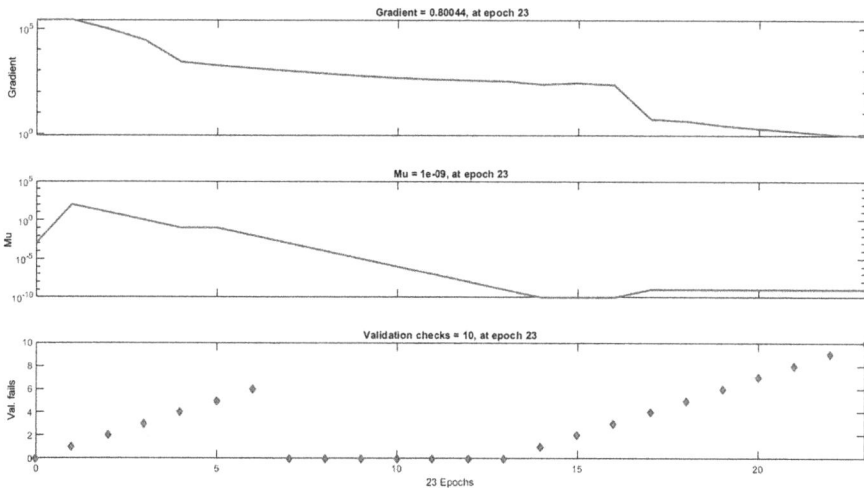

FIGURE 8.7 Training statistics for prediction by HLANN.

The accuracy of the proposed model during training and testing phases is shown in Figure 8.8. It is observed that the accuracy of the training and test phases are 100% and 96.9%, respectively.

Similarly, the accuracy of SVM and GPR models is explored and the training statistics of SVM-linear, SVM-quadratic, SVM-cubic, SVM-medium Gaussian, GPR-rational quadratic and GPR-squared exponential are shown in Table 8.4. The levels of accuracy of these models used for prediction are depicted in Figure 8.9a, b, c, d, e and f, respectively. The tests of prediction are carried out on regression learner toolbox in MATLAB. In this work, root-mean-square error (RMSE), R-squared and mean squared error (MSE) are taken to observe the prediction error, variance and average of the squares of the errors, respectively, as shown in Table 8.4.

It is observed from Table 8.4 and Figure 8.9 that the SVM model gives optimal accuracy when quadratic kernel method is selected, while in the case of GPR model, the highest accuracy is achieved on both the methods, i.e. rational quadratic and squared exponential kernel methods. Therefore, Q-SVM (quadratic SVM) and RQ-GPR (rational quadratic GPR) models are selected to compare the results with the proposed model.

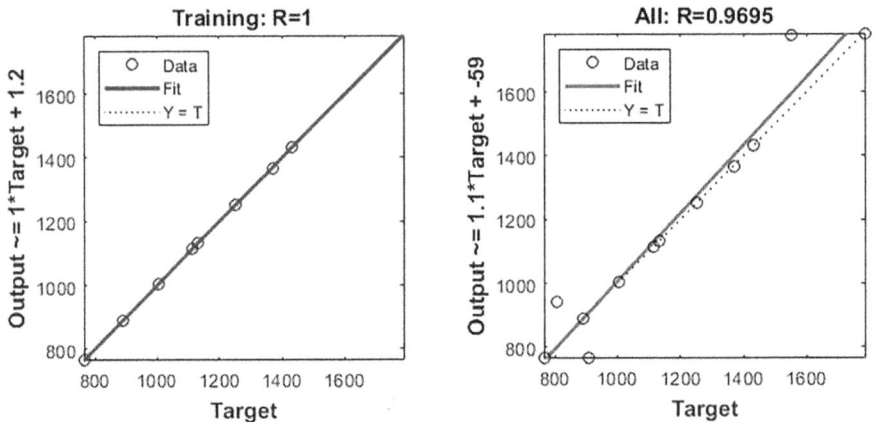

FIGURE 8.8 Accuracy of HLANN model during training and testing.

TABLE 8.4
Training Parameters for SVM and GPR Models

Sr. No.	Model	Kernel Method	Model Number	RMSE	R-squared	MSE
1	SVM	Linear	Model 1	78.088	0.95	6097.7
		Quadratic	Model 2	49.118	0.98	2412.6
		Cubic	Model 3	67.661	0.96	4578
		Medium Gaussian	Model 4	145.7	0.82	21229
2	GPR	Rational quadratic	Model 5	7.1461	1.00	51.067
		Squared exponential	Model 6	7.1461	1.00	51.067

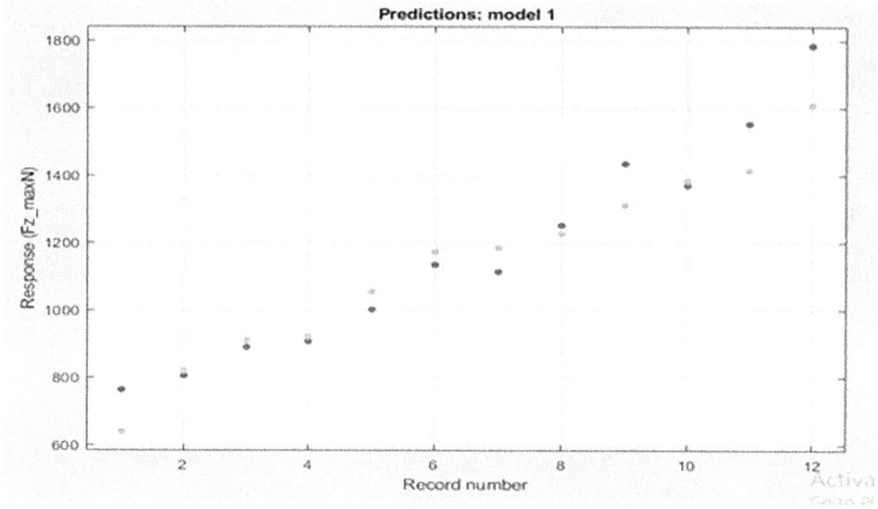

FIGURE 8.9A   Prediction accuracy of SVM-linear model.

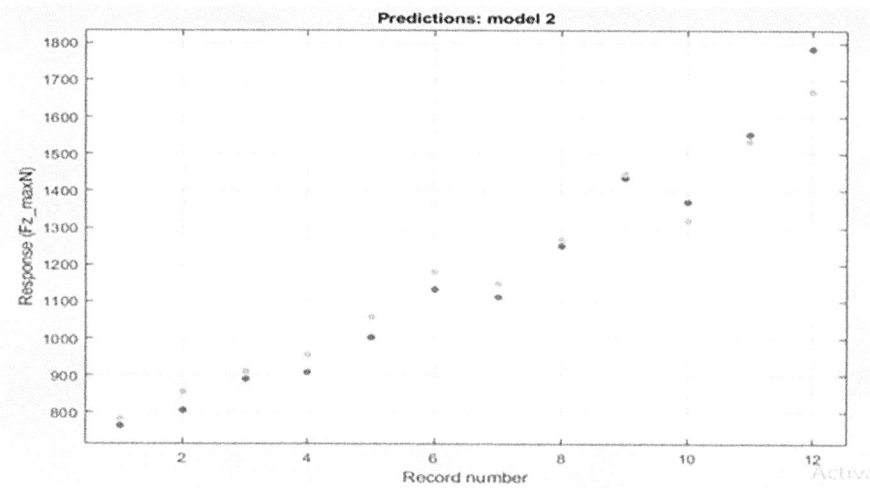

FIGURE 8.9B   Prediction accuracy of SVM-quadratic model.

8.3.4   COMPARISON OF THE ESTIMATED AND EXPERIMENTAL VALUES OF AXIAL FORCES

Figure 8.10 shows a comparison between the experimental and predicted values of axial peak forces using HLANN, SVM and GPR models, which has been delineated to depict the ability of prediction AI models. It can be observed from Figure 8.10 that the proposed model (HLANN) outperforms the SVM and GPR regression models.

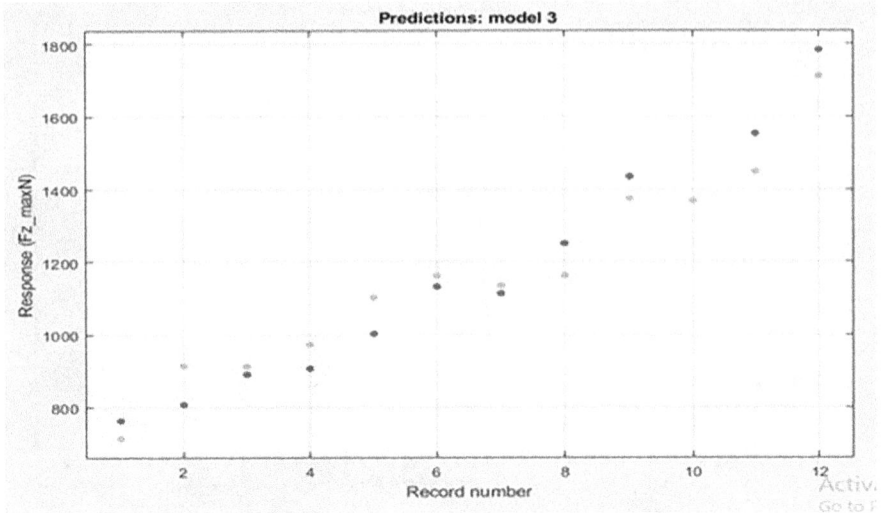

FIGURE 8.9C   Prediction accuracy of SVM-cubic model.

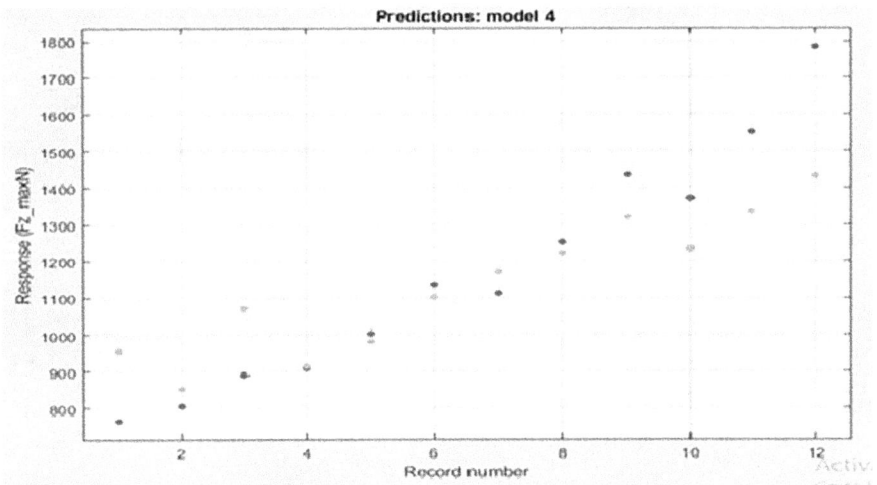

FIGURE 8.9D   Prediction accuracy of SVM-medium Gaussian model.

The results estimated by HLANN model are close to the experimental results, which shows the suitability and effectiveness of the proposed model. In addition, the errors between the estimated results by SVM model and the experimental results were greater than those produced between the estimated results by GPR model and the experimental results.

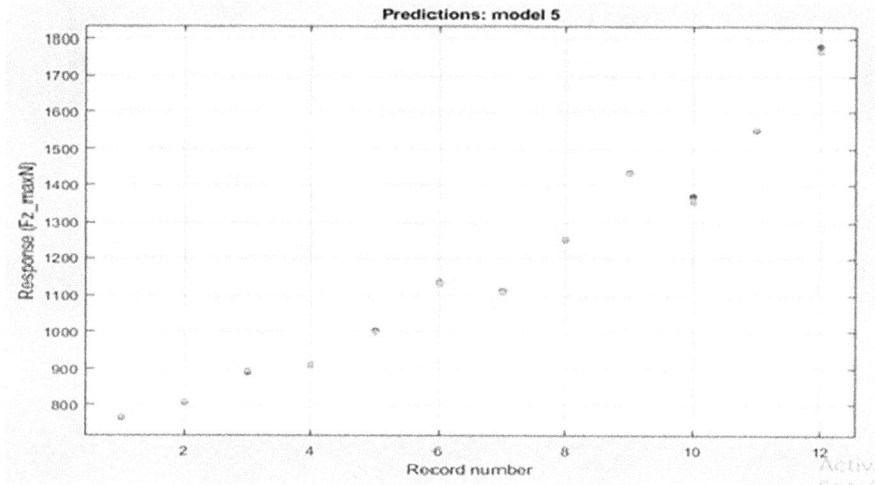

FIGURE 8.9E Prediction accuracy of GPR-rational quadratic model.

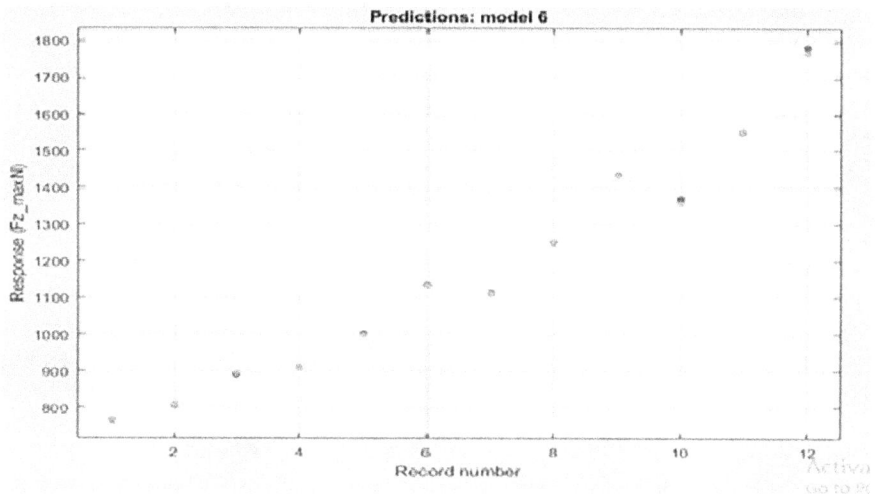

FIGURE 8.9F Prediction accuracy of GPR-squared exponential model.

8.4 CONCLUSIONS

The objective of this study was to investigate and predict the maximal axial forming forces required to produce conical frustums during SPIF technique by using a helical tool path. The input variables tool diameter and wall angle have been investigated to check their effects on maximal axial forces. The effects of interactions of these

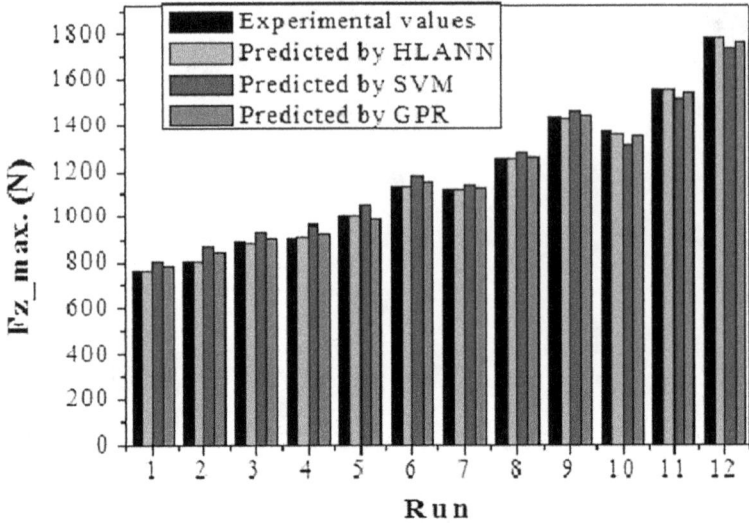

FIGURE 8.10   Comparison between the experimental and estimated values of axial peak forces.

factors have also been discussed. A hybrid learning-based ANN (HLANN) model was also proposed to predict the forming force in SPIF. The results of the HLANN model were compared with the experimental results and the results of two state-of-the-art regression models SVM and GPR. The comparison showed that the HLANN model outperforms the other regression methods. The accuracy of results predicted using HLANN model was found to be 96.90%. In the current work, the proposed model was trained and tested on a data set of small size.

It was observed from experimental campaign that the investigated factors were of great significance that can affect and control the process. When the combination of higher levels of tool diameter and feed rate is employed, the forming force increases very rapidly and, hence, becomes the limiting factor of forming tool and machinery that should obviously be avoided. The combination of lower tool diameter and higher wall angle can deform the material by lower forming forces at the expense of formability. Future work would target the investigation of effects of interactions of input variables on formability and surface roughness of formed components. In addition, a large experimental data set will be used to increase the accuracy of the proposed model.

REFERENCES

1. Ajay, and R.K. Mittal. *Incremental Sheet Forming Technologies: Principles, Merits, Limitations, and Applications.* CRC Press, Taylor & Francis, ISBN: 978-0-367-27674-4 (2020).
2. Kumar, A., and V. Gulati. Experimental investigation and optimization of surface roughness in negative incremental forming. *Measurement* 131(2019): 419–430.
3. Kumar, A., V. Gulati, and P. Kumar. Investigation of surface roughness in incremental sheet forming. *Procedia Computer Science* 133(2018): 1014–1020.

4. Kumar, A., V. Gulati, and P. Kumar. Effects of process parameters on surface roughness in incremental sheet forming. *Materials Today: Proceedings* 5, no. 14 (2018): 28026–28032.
5. Jeswiet, J., F. Micari, G. Hirt, A. Bramley, J. Duflou, and J. Allwood. Asymmetric single point incremental forming of sheet metal. *CIRP Annals* 54, no. 2 (2005): 88–114.
6. Bagudanch, I., G. Centeno, C. Vallellano, and M.L. Garcia-Romeu. Towards the Manufacturing of Near Net Shape Medical Prostheses in Polymeric Sheet by Incremental Sheet Forming. In: Gupta, K. (eds.) *Near Net Shape Manufacturing Processes*, pp. 1–33. Springer, Cham (2019).
7. Gulati, V., and A. Kumar. Investigation of some process parameters on forming force in single point incremental forming. *International Research Journal of Engineering and Technology* 4, no.4 (2017): 784–791.
8. Manju, V. Gulati, and A. Kumar. Process parameters and thickness reduction in single point incremental forming. *International Research Journal of Engineering and Technology* 7, no.1 (2020): 473–476.
9. Centeno, G., and I. Bagudanch, A.J. Martínez-Donaire, M.L. Garcia-Romeu, and C. Vallellano. Critical analysis of necking and fracture limit strains and forming forces in single-point incremental forming. *Materials & Design* 63 (2014): 20–29.
10. Suresh, K., S.D. Bagade, and S.P. Regalla. Deformation behavior of extra deep drawing steel in single-point incremental forming. *Materials and Manufacturing Processes* 30, no. 10 (2015): 1202–1209.
11. Kumar, A., V. Gulati, P. Kumar, V. Singh, B. Kumar, and H. Singh. Parametric effects on formability of AA2024-O aluminum alloy sheets in single point incremental forming. *Journal of Materials Research and Technology* 8, no. 1 (2019): 1461–1469.
12. Kumar, A., V. Gulati, P. Kumar, H. Singh, V. Singh, S. Kumar, and A. Haleem. Parametric investigation of forming forces in single point incremental forming. In: *International Conference on "Advances in Materials and Manufacturing Applications (IConAMMA 2018)"*, Bengaluru, India, during 16th–18th August, 2018.
13. Kumar, A., V. Gulati, P. Kumar, and H. Singh. Forming force in incremental sheet forming: a comparative analysis of the state of the art. *Journal of the Brazilian Society of Mechanical Sciences and Engineering* 41, no. 6 (2019): 251.
14. Kumar, A., V. Gulati, and P. Kumar. Investigation of process variables on forming forces in incremental sheet forming. *International Journal of Engineering & Technology* 10(2018): 680–684.
15. Kumar, A., V. Gulati, and P. Kumar. Experimental Investigation of Forming Forces in Single Point Incremental Forming. In: Shanker, K., Shankar, R., Sindhwani, R. (eds.) *Advances in Industrial and Production Engineering*, pp. 423–430. Springer, Singapore, 2019.
16. Uheida, E.H., G.A. Oosthuizen, and D. Dimitrov. Investigating the impact of tool velocity on the process conditions in incremental forming of titanium sheets. *Procedia Manufacturing*, 7 (2017): 345–350.
17. Alsamhan, A., A.E. Ragab, A. Dabwan, M.M. Nasr, and L. Hidri. Prediction of formation force during single-point incremental sheet metal forming using artificial intelligence techniques. *PloS one* 14, no. 8 (2019): e0221341.
18. Sakhtemanian, M.R., M. Honarpisheh, and S. Amini. A novel material modeling technique in the single-point incremental forming assisted by the ultrasonic vibration of low carbon steel/commercially pure titanium bimetal sheet. *The International Journal of Advanced Manufacturing Technology* 102, no. 1–4 (2019): 473–486.
19. Long, Y., Y. Li, J. Sun, I. Ille, J. Li, and J. Twiefel. Effects of process parameters on force reduction and temperature variation during ultrasonic assisted incremental sheet forming process. *The International Journal of Advanced Manufacturing Technology* 97, no. 1–4 (2018): 13–24.

20. Honarpisheh, M., M. Keimasi, and I. Alinaghian. Numerical and experimental study on incremental forming process of Al/Cu bimetals: Influence of process parameters on the forming force, dimensional accuracy and thickness variations. *Journal of Mechanics of Materials and Structures* 13, no. 1 (2018): 35–51.

21. Zhai, W., Y. Li, Z. Cheng, L. Sun, F. Li, and J. Li. Investigation on the forming force and surface quality during ultrasonic-assisted incremental sheet forming process. *The International Journal of Advanced Manufacturing Technology* 106 (2020): 1–17.

22. Kumar, A., and V. Gulati. Experimental investigations and optimization of forming force in incremental sheet forming. *Sādhanā* 43, no. 10 (2018): 159.

23. Kumar, A., and V. Gulati. Optimization and investigation of process parameters in single point incremental forming. *Indian Journal of Engineering & Materials Sciences* 27, no. 2, (2020); 246–255. (Accepted, Article in Press).

24. Chang, Z., M. Li, and J. Chen. Analytical modeling and experimental validation of the forming force in several typical incremental sheet forming processes. *International Journal of Machine Tools and Manufacture* 140 (2019): 62–76.

25. Oleksik, V., A. Pascu, A. Gavrus, and M. Oleksik. Experimental studies regarding the single point incremental forming process. *Academic Journal of Manufacturing Engineering* 8 (2010):51–56.

26. Fiorentino, A., E. Ceretti, A. Attanasio, L. Mazzoni, and C. Giardini. Analysis of forces, accuracy and formability in positive die sheet incremental forming. *International Journal of Material Forming* 2, no. 1 (2009): 805.

27. Jain, A.K., J. Mao, and K.M. Mohiuddin. Artificial neural networks: A tutorial. *Computer* 29, no. 3 (1996): 31–44.

28. Anwar, S., M.M. Nasr, M. Alkahtani, and A. Altamimi. Predicting surface roughness and exit chipping size in BK7 glass during rotary ultrasonic machining by adaptive neuro-fuzzy inference system (ANFIS). In *Proceedings of the International Conference on Industrial Engineering and Operations Management*, Rabat, Morocco, 2017.

29. Kurra, S., N.H. Rahman, S.P. Regalla, and A.K. Gupta. Modeling and optimization of surface roughness in single point incremental forming process. *Journal of Materials Research and Technology* 4, no. 3 (2015): 304–313.

30. Do, V.-C., and Y.-S. Kim. Effect of hole lancing on the forming characteristic of single point incremental forming. *Procedia Engineering* 184(2017): 35–42.

31. Khan, M.S., F. Coenen, C. Dixon, S. El-Salhi, M. Penalva, and A. Rivero. An intelligent process model: predicting springback in single point incremental forming. *The International Journal of Advanced Manufacturing Technology* 76, no. 9–12 (2015): 2071–2082.

32. Zhang, N., J. Xiong, J. Zhong, and K. Leatham. Gaussian process regression method for classification for high-dimensional data with limited samples. In *2018 Eighth International Conference on Information Science and Technology (ICIST)*, Cordoba, Granada, and Seville, Spain, pp. 358–363. IEEE, 2018.

33. Noble, W.S. What is a support vector machine? *Nature Biotechnology* 24, no. 12 (2006): 1565–1567.

34. Min, J.H., and Y.-C. Lee. Bankruptcy prediction using support vector machine with optimal choice of kernel function parameters. *Expert Systems with Applications* 28, no. 4 (2005): 603–614.

35. Quinonero-Candela, J., C.E. Rasmussen, and C.K. Williams. Approximation methods for Gaussian process regression. In: Bottou, L., Chapelle, O., DeCoste, D., Weston, J. (eds.) *Large-Scale Kernel Machines*, pp. 203–223. MIT Press, Cambridge, MA, 2007.

9 Development of Non-dominated Genetic Algorithm Interface for Parameter Optimization of Selected Electrochemical-Based Machining Processes

D. Singh and R. S. Shukla
Sardar Vallabhbhai National Institute of Technology

CONTENTS

9.1   INTRODUCTION

In the last decade, the usage of advanced machining processes (AMP) has rapidly grown due to the increase in the use of hard, high-strength, and temperature-resistant materials in engineering industries. Materials such as nitralloys, Nimonic alloys, Hastelloy, and carbides are difficult for machining, and these materials have found their wide applications in special cutting tools, aircrafts, turbines, etc. These products require high precision in machining and manufacturing. It is difficult to machine these materials with the traditional machining processes due to their excellent strength. These difficulties can be overcome with the usage of a special type of advanced machining process. In this chapter, AMP such as electrochemical micromachining (EMM), electrochemical machining (ECM), and electrochemical turning (ECT) are considered, for which the non-dominated optimal set of process parameters are obtained to enhance the performance characteristics. To accomplish the desired objective, a graphical user interface (GUI) that mimics a metaheuristic algorithm, i.e., non-dominated sorting genetic algorithm (NSGA-II), is proposed.

The application of the ECM-based processes has increased in last decade, and researchers have attempted several works related to the considered process. Kuriakose et al. (2005) attempted an experimental investigation on a variant of electric discharge machining (i.e., wire EDM). The data obtained were modeled using multiple regression modeling, and NSGA-II was applied to this modeling to get a non-dominated set of optimal solution. Konak et al. (2006) presented a different multi-objective optimization to promote solution diversity. Jain and Jain (2007) attempted parameter optimization of the ECM process, i.e., tool feed rate, electrolyte flow velocity, and applied voltage, with an objective to minimize geometrical inaccuracy subject to temperature, choking, and passivity constraints using real-coded GA optimization technique. Asokan et al. (2008) developed ANN-based multiple regression models to determine the optimal ECM parameters (i.e., current, voltage, flow rate, and gap). Analysis of variance (ANOVA) was used to determine the significant parameters that affect performance parameters (i.e., MRR and Ra). Yusoff et al. (2011) reviewed several multi-objective optimization techniques with their brief application and suggested that unlike the single-objective optimization technique, the multi-objective optimization simultaneously optimizes each objective without affecting the other solutions.

Teimouri and Sohrabpoor (2013) investigated the effects of ECM process parameters (i.e., electrolyte concentration, electrolyte flow rate, applied voltage, and feed rate) on MRR and Ra using an adaptive neuro-fuzzy inference system (ANFIS) based on the predictive models for experimental observations. They attempted performance parameter optimization using the cuckoo optimization algorithm, and the results were confirmed using the confirmatory tests. Acharya et al. (2013) studied the effects of four process parameters, i.e., current, voltage, flow rate of electrolyte, and inter-electrode gap, on the performance parameters MRR and Ra during machining of cylindrical blanks of 20 mm diameter and 40 mm height, which are made of hardened steel. They used NSGA to obtain the optimal process parameters to maximize MRR and minimize Ra simultaneously. Ayyappan and Sivakumar (2015) studied the microstructure of the surfaces of the material steel 20MnCr5 obtained during machining using ECM to control the influence of electrolyte on the machining surfaces. Sohrabpoor et al. (2015) conducted an experimental investigation on the ECM process and used teaching cuckoo optimization algorithm (TCOA) to obtain the optimum process parameters.

It is observed from the literature that the researchers have attempted parameter optimization using metaheuristic algorithms, experimental investigation, and microstructural analysis to investigate the effects of ECM process parameters on the performance of the process. Further, it is observed from the literature that the multi-objective optimization has its applications in building designing (Wang et al., 2005), scheduling problems (Pasupathy et al., 2006), power planning (Ramesh et al., 2012), energy conservation systems (Yang et al., 2016), and sustainable designs in energy supply (Majewski et al., 2017). This shows the effectiveness of the multi-objective optimization in various domains, and thus, in the present study, a multi-objective optimization-based GUI is proposed. After the detailed study of the literature, a motivation is created in the researchers' mind to extend the application of the considered algorithm by reshaping it in the GUI form. As far as the authors' knowledge is concerned, only few attempts have been made to develop GUI for use in the field parameter optimization of advanced machining processes. In this work, a GUI is proposed based on a metaheuristic technique, i.e., NSGA-II. To see the effectiveness of the NSGA-II-based GUI, it is applied to selected case studies of advanced machining processes, namely electrochemical micromachining (EMM), electrochemical machining (ECM), and electrochemical turning (ECT).

The details of the NSGA-II-based GUI are described in the next section. In Section 9.3, the parameter optimization for the three considered machining processes, ECM, EMM, and ECT, is attempted by using the developed GUI.

9.2 METHODOLOGY

This section reports the developed GUI for the considered multi-objective optimization technique. The developed NSGA-II-based GUI is user-friendly due to the ease in operating the interface for the parameter optimization problems.

9.2.1 Non-dominated Sorting Genetic Algorithm – Graphical User Interface (NSGA-GUI)

Non-dominated sorting genetic algorithm (NSGA-II) was developed by Kalyanmoy Deb in 2002. In NSGA, a random population, i.e., parent population, is generated. The GA is a population-based algorithm that uses genetic operators (i.e., crossover and mutation) to search out the optimal solution (Deb et al., 2002). NSGA-II is a multi-objective optimization technique that uses the genetic operators similar to a GA for determining the set of non-dominated optimal values. In NSGA-II, a shared fitness value is assessed for the selection criterion. The shared fitness value is computed based on the ranking of the solution and crowding distance of the considered objectives. The procedure of NSGA-II is briefly explained below.

Chromosomes are generated randomly, and to obtain the non-domination level, each chromosome solution is compared with other solutions present in the same level. Then, it is verified whether the obtained solution obeys the rules given in Eq. (9.1) with respect to all other solutions present in chromosome population.

$$\text{Obj. } 1[i] > \text{Obj.1}[j] \text{ and Obj.2}[i] \geq \text{Obj.2}[j] \text{ or Obj.1}[i] \geq \text{Obj.1}[j] \text{ and}$$
$$\text{Obj. } 2[i] > \text{Obj.2}[j], i \neq j \tag{9.1}$$

If the rules given in Eq. (9.1) are satisfied for all the chromosomes present in the front while comparing, then it is marked as non-dominated; otherwise, it is marked as dominated. This same process is repeated for all the chromosomes to classify the dominated and non-dominated solutions. These non-dominated solutions obtained are equally good compared to other values present in the same front. All the non-dominated chromosomes (*n*) in the first sorting are ranked 1. For the remaining chromosomes (*N-n*), with respect to its non-domination level, ranks are assigned. The solutions present in the same front are assigned a rank that is equal to their non-domination level. If solutions 1 and 2 are compared and it is observed that solution 1 is better in both the objectives (i.e., Obj.1 and Obj.2) than solution 2, then solution 1 dominates solution 2, or in other words, solution 1 is non-dominated by solution 2. The subpopulation with rank 1 sets of solutions in the first front is assigned a dummy fitness F_1 (Senthilkumar et al., 2011; Deb et al., 2002, Konak et al., 2006; Kuriakose and Shunmugam, 2005; Srinivas and Deb, 1995; Yusoff et al., 2011). The "normalized Euclidean distance" (d_{ij}) between chromosomes present in the front set is calculated using Eq. (9.2).

$$d_{ij} = \sqrt{\sum_{x=1}^{nvar} \left(\frac{x_d^{(i)} - x_d^{(j)}}{x_d^{max} - x_d^{min}} \right)^2} \qquad (9.2)$$

where x_d is the value of the *d*th decision variable, *nvar* is the number of variables, and *i* and *j* are chromosomes numbers.

Then, the sharing function value for each chromosome present in the first front set is computed using Eq. (9.3).

$$sh(d_{ij}) = \begin{cases} 1 - \left(\dfrac{d_{ij}}{\sigma_{share}} \right)^2, & \text{if } d_{ij} < \sigma_{share} \\ 0, & \text{otherwise} \end{cases} \qquad (9.3)$$

where σ_{share} is the distance permitted between any two chromosomes (Kuriakose and Shunmugam, 2005). A niche count (nc_i) that provides an estimate of the extent of crowding near a chromosome is calculated using Eq. (9.4).

$$nc_i = \sum_{j=1}^{N} sh(d_{ij}) \qquad (9.4)$$

where *N* is the total population. Then, the shared fitness values (*F*) are calculated using Eq. (9.5).

$$F = \frac{F_1}{nc_i} \qquad (9.5)$$

After calculating the shared fitness, a smaller value is deducted from the shared fitness to assign dummy fitness (F_2) to the subsequent front with rank 2, and the steps are repeated (Kuriakose and Shunmugam, 2005). Based on the shared fitness value, chromosomes selected depending on the values obtained for cumulative probability.

Genetic operators are applied to these selected chromosomes for the calculation of the objective function values (Deb et al., 2002; Kuriakose and Shunmugam, 2005). The flowchart of NSGA-II to be used in GUI is shown in Figure 9.1.

The coding of NSGA-II algorithm in programming software is a little complex task for each individual researcher. As observed from the literature, the considered

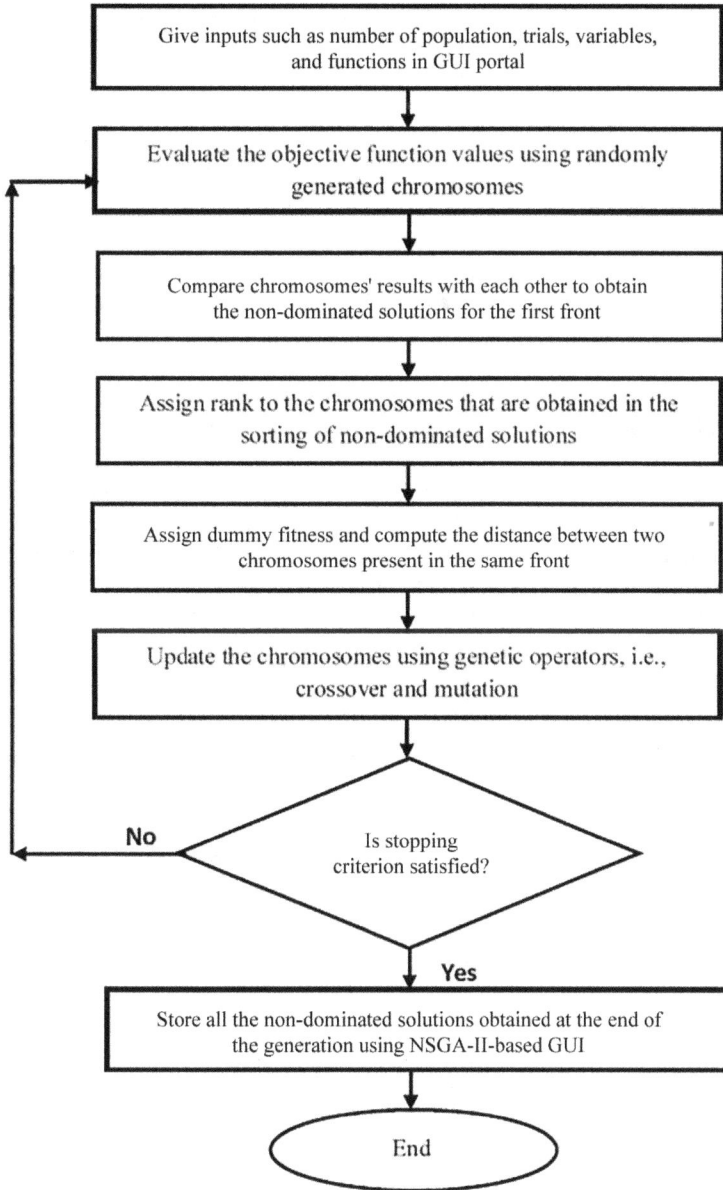

FIGURE 9.1   Flowchart of NSGA-II used in GUI.

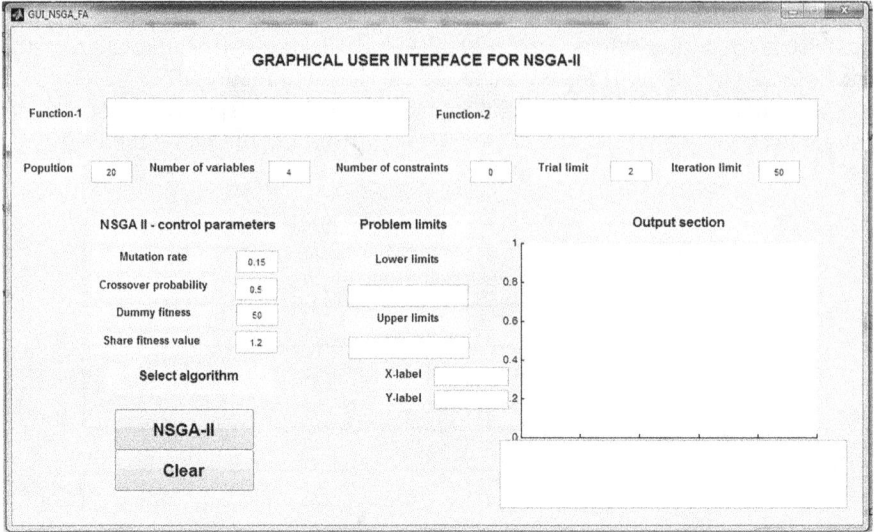

FIGURE 9.2   Graphical user interface for NSGA-II.

algorithm has a wide range of applications in problem-solving of optimization problems. Therefore, to ease the use of this algorithm, a GUI is proposed as an end display for the user to execute the specified task of parameter optimization in continuous domain. The characteristic "user-friendly" is significant as it decides the reachability and maximizes the use of interface among the users. The proposed interface has tried to accomplish the considered characteristics. The proposed GUI is developed using the features of MATLAB GUIDE environment (Chapman, 2008; Hahn and Valentine, 2010). In the present research, an attempt is made to connect a metaheuristic technique, i.e., NSGA-II, in a single interface as shown in Figure 9.2. With the use of this interface, users can deal with optimization problems without bothering about the mathematical programming of the algorithm. They can concentrate on the solution to the optimization problems.

9.3   APPLICATIONS OF NSGA-GUI IN ADVANCED MACHINING PROCESSES

This section demonstrates the simulation results of the developed GUI based on a metaheuristic technique, i.e., NSGA-II, to obtain the alternate non-dominated solutions of the operating parameters for the considered electrochemical-based processes.

9.3.1   ELECTROCHEMICAL MACHINING (ECM)

The present case study is taken from the research paper by Bhattacharyya and Sorkhel (1998). They investigated the ECM process parameters by conducting experimentations on an ECM setup. They selected a range of process parameters during the experimentation as follows: "electrolyte concentration" as 15–90 (g/l), "electrolyte flow rate" as 10–14 (l/min), "applied voltage" as 10–40 (V), and "inter-electrode gap" as 0.4–1.2 (mm). A cylindrical workpiece made of "EN-8 steel" was machined using a brass tool to obtain the influence of the considered process parameters on MRR (g/min) and overcut (mm). The parametric levels of the process parameters for given coded values (X_i) can be uncoded to the actual variable (X) using Eq. (9.6). A central composite rotatable second-order experimentation plan was used to develop regression models for MRR and overcut as given in Eqs. (9.7)–(9.8), respectively. These same regression models are considered in the present study to obtain the alternative non-dominated solutions using NSGA-II-based GUI.

$$\text{coded value } (X_i) = \frac{2X - (X_{max} + X_{min})}{\dfrac{X_{max} - X_{min}}{2}} \tag{9.6}$$

where the coded values for X_i are -2, -1, 0, 1, and 2, and X_{max} and X_{min} are the maximum and minimum values of the actual variable.

$$MRR = 0.6244 + 0.1523x_1 + 0.0404x_2 + 0.1519x_3 - 0.1169x_4 + 0.0016x_1^2$$

$$+ 0.0227x_2^2 + 0.0176x_3^2 - 0.0041x_4^2 + 0.0077x_1x_2 + 0.0119x_1x_3$$

$$- 0.0203x_1x_4 + 0.0103x_2x_3 - 0.0095x_2x_4 + 0.0300x_3x_4 \tag{9.7}$$

$$OC = 0.3228 + 0.0214x_1 - 0.0052x_2 + 0.0164x_3 + 0.0118x_4 - 0.0041x_1^2$$

$$- 0.0122x_2^2 + 0.0027x_3^2 + 0.0034x_4^2 - 0.0059x_1x_2 - 0.0046x_1x_3 - 0.0059x_1x_4$$

$$+ 0.0021x_2x_3 - 0.0053x_2x_4 - 0.0078x_3x_4 \tag{9.8}$$

The selected control parameters are a number of iterations of 100, a sharing fitness of 1.2, a population size of 200, and a dummy fitness of 50 for obtaining the alternative solutions in the considered ECM process. These control parameters are selected on the basis of trail runs for the given problem. The aim is to maximize the MRR and minimize the overcut simultaneously. Since the objectives are conflicting in nature, the first objective (MRR) is modified and converted into a minimization problem. The objectives are given in Eq. (9.9).

$$\text{Objective } 1 = -(MRR) \qquad \text{Objective } 2 = \text{overcut} \tag{9.9}$$

Initially, using control parameters of NSGA-II, the values of the objective function are obtained. Twenty-four solutions are obtained at the end of 100 generations with a computation time of 451.73 s. The computer processor used is Dell-Intel I3 fourth generation with 2GB RAM. The Pareto-optimal solutions along with the performance parameters and actual values of process parameters are given in Table 9.1. Figure 9.3 shows the non-dominated solutions termed as a Pareto-optimal front.

TABLE 9.1

Pareto-Optimal Solutions for the ECM Process

No.	Electrolyte Concentration (g/l)		Electrolyte Flow Rate (l/min)		Applied Voltage (V)		Inter-Electrode Gap (mm)		MRR (g/min)	Overcut (mm)
	Coded	Actual	Coded	Actual	Coded	Actual	Coded	Actual		
1	1.9335	74.0029	1.9765	13.9765	0.5845	22.9227	-0.0879	0.7824	1.2567	0.2774
2	-0.9873	30.1906	1.9804	13.9804	-1.4838	12.5806	-1.3118	0.5376	0.6331	0.2010
3	1.8045	72.0675	1.9765	13.9765	-0.4711	17.6441	-1.9178	0.4164	1.3555	0.2872
4	-1.7302	19.0469	1.9921	13.9921	-0.3343	18.3284	-1.8631	0.4273	0.6530	0.2114
5	1.4956	67.4340	1.9608	13.9609	1.8670	29.3352	-0.2170	0.7565	1.4843	0.3110
6	-0.0059	44.9120	1.5112	13.5112	-1.9061	10.4692	-1.9374	0.4125	0.8306	0.2359
7	-1.5347	21.9795	-1.8475	10.1524	-1.9609	10.1955	-1.5659	0.4868	0.4509	0.1399
8	1.7459	71.1877	1.9843	13.9843	-0.5024	17.4877	-1.5620	0.4875	1.2797	0.2795
9	1.8788	73.1818	2.0000	14.0000	0.7409	23.7047	0.1584	0.8316	1.2466	0.2751
10	1.8436	72.6539	1.9374	13.9374	1.0146	25.0733	-0.4007	0.7198	1.3727	0.2937
11	-0.2757	40.8651	1.7458	13.7458	-1.2922	13.5386	-1.8357	0.4328	0.8247	0.2353
12	1.9805	74.7067	1.8709	13.8709	1.6637	28.3186	-1.9178	0.4164	1.7131	0.3571
13	1.5855	68.7829	1.9491	13.9491	-1.8045	10.9775	-1.6793	0.4641	1.1324	0.2558
14	1.2454	63.6804	1.9726	13.9726	-1.5425	12.2873	-1.1554	0.5689	0.9907	0.2520
15	-1.7263	19.1056	-1.9374	10.0625	-0.6041	16.9794	-1.5659	0.4868	0.4557	0.1573
16	-1.5503	21.7449	1.9022	13.9022	-1.6441	11.7790	-1.5073	0.4985	0.5454	0.1771
17	-1.8983	16.5249	-1.8241	10.1759	-1.5268	12.3655	-1.8436	0.4312	0.4248	0.1190

(Continued)

TABLE 9.1 (Continued)
Pareto-Optimal Solutions for the ECM Process

No.	Electrolyte Concentration (g/l)		Electrolyte Flow Rate (l/min)		Applied Voltage (V)		Inter-Electrode Gap (mm)		MRR (g/min)	Overcut (mm)
	Coded	Actual	Coded	Actual	Coded	Actual	Coded	Actual		
18	0.8387	57.5807	1.9961	13.9960	-0.5180	17.4095	-1.8436	0.4312	1.1440	0.2720
19	-1.4369	23.4457	1.7693	13.7693	-1.3587	13.2062	-1.8670	0.4265	0.6133	0.1915
20	1.9296	73.9442	1.9569	13.9569	1.65982	28.2991	-1.5855	0.4828	1.6791	0.3412
21	-1.8241	17.6393	1.3001	13.3001	-1.9257	10.3714	-1.9491	0.4101	0.4901	0.1668
22	1.6481	69.7214	1.9960	13.9960	1.7028	28.5141	-0.1661	0.7667	1.4738	0.3022
23	-0.2678	40.9824	1.9023	13.9022	-1.5777	12.1114	-1.4095	0.5180	0.7582	0.2231
24	-1.9844	15.2346	-1.8123	10.1876	-1.9804	10.0977	-1.4447	0.5110	0.3725	0.1168

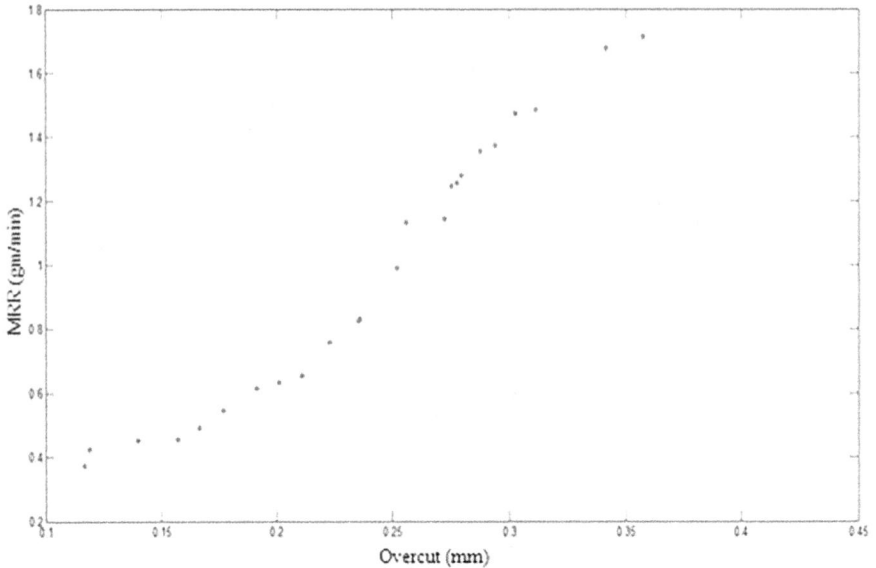

FIGURE 9.3 Solution set and Pareto-optimal front for the ECM process.

The solutions obtained for the ECM process using NSGA-II cannot be compared, as Bhattacharyya and Sorkhel (1998) did not consider the problem of multi-objective optimization of the objectives. However, these non-dominated solutions obtained using NSGA-II are found comparable with the experimental results obtained by Bhattacharyya and Sorkhel (1998). Furthermore, Mukherjee and Chakroborty (2013) considered weighted-based multi-objective optimization and obtained three solutions by giving different weights to the objectives. The solutions obtained for the performance parameter {(MRR, overcut)} were {(1.3230, 0.2290), (0.8186, 0.1896), (1.4489, 0.2346)} for different cases with different weights to the considered objectives. The solutions obtained using NSGA-II are also found better compared to the solutions obtained using weighted-based multi-objective optimization considered by Mukherjee and Chakroborty (2013). Multiple sets of solutions are obtained, so, according to the requirements, one can choose the appropriate solution set. The condition of the Pareto front is an outcome of the incessant nature of the given problem considered for optimization. The results given in Table 9.1 clearly show the 24 non-dominated solutions for the considered ECM process. The range of input process parameters is reflected, and no partiality toward the lower bound values or higher bound values is seen. This is the main advantage of NSGA-II. The objectives are conflicting in nature, and it is clearly observed in the trends obtained using NSGA-II.

9.3.2 ELECTROCHEMICAL MICROMACHINING (EMM)

The EMM process is utilized for the microfabrication of complex microstructures and film processing. The present case study is taken from Malapati and Bhattacharya (2011). They performed experimental investigations to analyze the influence of

EMM process parameters on MRR and machining accuracy factors such as width overcut (WOC), length overcut (LOC), and linearity. They selected a range of EMM process parameters during the experimentation as follows: "pulse frequency" as 40–80 (kHz), "machining voltage" as 7–11 (V), "duty ratio" as 30–70 (%), "electrolyte concentration" as 50–90 (g/l), and "microtool feed rate" as 150–250 (μm/s). They used copper plates as the workpiece material and Tungsten as the microtool material. They used a central composite half-fraction rotatable second-order design to perform 32 experimental trials and developed regression models for MRR, WOC, LOC, and linearity as given in Eqs. (9.10)–(9.13), *respectively*. In the present case study, the same setting of EMM process parameters and regression models are considered for the parameter optimization using NSGA-II-based GUI.

$$MRR = 2.59446 + 0.237004x_2 + 0.426029x_3 + 0.340938x_4 - 0.385790x_1^2$$

$$+ 0.295540x_2^2 - 0.299427x_3^2 + 0.224331x_1x_5 + 0.327156x_2x_3 - 0.243256x_2x_4$$

$$(9.10)$$

$$WOC = 0.124856 + 0.0288208x_4 + 0.0270208x_5 - 0.0173807x_3^2 + 0.0140443x_5^2$$

$$+ 0.0203563x_1x_4 - 0.0257063x_1x_5 - 0.0209313x_2x_3 \qquad (9.11)$$

$$LOC = 0.403467 + 0.0526542x_2 + 0.0604563x_1x_2 + 0.0457563x_1x_3 + 0.0344187x_1x_4$$

$$- 0.166544x_2x_4 - 0.0554562x_2x_5 + 0.0495813x_3x_4$$

$$+ 0.0413687x_3x_5 + 0.0534565x_4x_5 \qquad (9.12)$$

$$Linearity = 0.0721852 + 0.00868750x_1 + 0.0112125x_2 + 0.0171458x_4$$

$$- 0.0128688x_1x_3 + 0.0151188x_1x_4 + 0.00841875x_2x_4$$

$$- 0.0116062x_2x_5 - 0.00749375x_4x_5 \qquad (9.13)$$

In multi-objective optimization of the EMM process, the objectives (MRR, WOC, LOC, and linearity) are optimized simultaneously in the GUI using NSGA-II. The objective functions are taken from Malapati and Bhattacharyya (2011). Similar to the ECM process, the objective (MRR) is modified and converted into a minimization problem. The objectives are given in Eq. (9.14).

$$Objective\, 1 = -(MRR)$$

$$Objective\, 2 = LOC$$

$$subject\, to \qquad\qquad (9.14)$$

$$WOC \leq 0.0214$$

$$Linearity \leq 0.0328$$

where 0.0214 and 0.0328 are the minimum values of WOC and linearity obtained from the experimental data of Malapati and Bhattacharyya (2011). A set of two solutions is found at the end of 500 generations with a computation time of 224.38 s. The Pareto-optimal solutions along with the performance parameters and actual values of process parameters are reported in Table 9.2. The solution obtained for the EMM process using NSGA-II cannot be compared, as Malapati and Bhattacharyya (2011) did not consider the problem of multi-objective optimization for the considered objectives. However, these non-dominated solutions obtained using NSGA-II-based GUI are found comparable with the single-objective optimization results of Malapati and Bhattacharyya (2011). The objectives are conflicting in nature, and it is clearly observed from the values obtained using NSGA-II. The constraint values of WOC and linearity are satisfied with the process parameters that are obtained using NSGA-II-based GUI. Thus, the solution obtained using the NSGA-II-based GUI is satisfying the constraints WOC and linearity considered in the EMM process.

9.3.3 ELECTROCHEMICAL TURNING (ECT)

The present case study is taken from El-Taweel and Gouda (2011). It is one of the variants of the ECM process and is termed as "electrochemical turning." It is utilized for finishing processes. They performed an experimental investigation to analyze the influence of ECT process parameters on MRR, Ra, and roundness error (RE). They selected a range of ECT process parameters during the experimentation as follows: "applied voltage" as 10 (V) to 40 (V), "wire feed rate" as 0.1–0.5 (mm/min), "wire diameter" as 0.2–2 (mm), "overlap distance" as 0.02–0.06 (mm), and "rotational speed" as 300–900 (rpm). They used a central composite second-order rotatable design to perform 32 experimental trials. The mathematical regression models are remodeled for the experimental values of El-Taweel and Gouda (2011) using Minitab software for the considered performance characteristics MRR, Ra, and RE as given in Eqs. (15)–(17), respectively. In the present case study, the same setting of the EMM process parameters and the remodeled regression models are considered for the parameter optimization using the NSGA-II-based GUI.

$$MRR = 0.214932 + 0.050667x_1 + 0.016583x_2 + 0.006167x_3 + 0.004667x_4$$

$$+ 0.010917x_5 + 0.002068x_1^2 - 0.005557x_2^2 - 0.021307x_3^2 - 0.004057x_4^2$$

$$+ 0.004068x_5^2 - 0.0005x_1x_2 + 0.004375x_1x_3 - 0.010625x_1x_4 + 0.004375x_1x_5$$

$$+ 0.014375x_2x_3 + 0.007625x_2x_4 + 0.011875x_2x_5 - 0.002x_3x_4 + 0.0135x_4x_5$$

$$(9.15)$$

TABLE 9.2

Pareto-Optimal Solutions for the EMM Process

Pulse Frequency (kHz)		Machining Voltage (V)		Duty Ratio (%)		Electrolyte Concentration (g/l)		Microtool Feed Rate (µm/s)		MRR (mg/min)	LOC (mm)	WOC (mm)	Linearity (mm)
Coded	Actual	Coded	Actual	Coded	Actual	Coded	Actual	Coded	Actual				
1.3743	73.7439	1.2219	10.2219	0.8974	58.9736	−1.2610	57.3900	0.3304	208.2600	3.1432	0.7342	0.0151	0.0196
1.2923	72.9228	0.2405	9.2405	1.8709	68.7097	−0.8269	61.7302	0.2483	206.2072	1.7588	0.4701	0.0083	0.0238

$$\mathrm{Ra} = 1.68193 + 0.16708x_1 + 0.06708x_2 - 0.20792x_3 + 0.12208x_4 - 0.13875x_5$$

$$- 0.02568x_1^2 - 0.01943x_2^2 - 0.01943x_3^2 - 0.01943x_4^2 - 0.03193x_5^2$$

$$+ 0.11188x_1x_2 + 0.02438x_1x_3 + 0.27938x_1x_4 + 0.03312x_1x_5 + 0.03062x_2x_3$$

$$+ 0.08563x_2x_4 - 0.01062x_2x_5 + 0.04813x_3x_4 + 0.15188x_3x_5 + 0.18188x_4x_5$$

$$\tag{9.16}$$

$$\mathrm{RE} = 7.21250 + 1.98917x_1 - 1.31000x_2 + 1.71x_3 - 1.09250x_4 - 2.39250x_5$$

$$+ 0.50000x_1^2 + 0.02125x_2^2 + 0.19375x_3^2 + 0.27875x_4^2 + 0.55875x_5^2 - 0.17375x_1x_2$$

$$- 0.31625x_1x_3 + 0.03250x_1x_5 - 0.41125x_2x_3 + 0.40250x_2x_4 - 0.005x_2x_5$$

$$- 0.06750x_3x_4 - 0.62250x_3x_5 + 0.91375x_4x_5$$

$$\tag{9.17}$$

In multi-objective optimization of the ECT process, the objectives (MRR, Ra, and RE) are optimized simultaneously in the GUI using NSGA-II. Here, the aim is to maximize the MRR and minimize the Ra subject to the constraint RE for the considered ECT machining process. The objective functions are taken from El-Taweel and Gouda (2011). Similar to the ECM and EMM problems, the objective (MRR) is converted into a minimization problem. The objectives are given in Eq. (9.18).

$$\text{Objective} 1 = -(\mathrm{MRR})$$

$$\text{Objective} 2 = \mathrm{Ra}$$

$$\text{subject to}$$

$$\tag{9.18}$$

$$\mathrm{RE} \le 3.5$$

where 3.5 is the minimum value of the RE taken from the experimental data of El-Taweel and Gouda (2011). The Pareto-optimal solutions along with the performance parameters and actual values of process parameters are reported. Four solutions are found at the end of 1000 generations with a computation time 576.43 s. The solution obtained for the ECT process using NSGA-II can be compared with the results of the considered case study, as El-Taweel and Gouda (2011) considered the problem of weighted-based multi-objective optimization of the objectives, and the results obtained for the parameters are as follows: MRR = 0.2980 g/min, Ra = 1.1294 μm, and RE = 5.5432 μm. Furthermore, Mukherjee and Chakroborty (2013) considered weighted-based multi-objective optimization using BBO by giving equal weights to the objectives and obtained the following values for the parameters: MRR = 0.3545 g/min, Ra = 0.8303μm, and RE = 2.0456 μm. However, the non-dominated solutions obtained using NSGA-II are better than those achieved by Mukherjee and Chakroborty (2013) and El-Taweel and Gouda (2011). The results reported in Table 9.3 clearly show the four non-dominated solutions for the considered ECT process.

TABLE 9.3
Pareto-Optimal Solutions for the ECT Process

Applied Voltage (V)		Wire Feed Rate (mm/min)		Wire Diameter (mm)		Overlap Distance (mm)		Rotational Speed (rpm)		MRR (g/min)	Ra (μm)	RE (μm)
Coded	Actual	Coded	Actual	Coded	Actual	Coded	Actual	Coded	Actual			
1.1672	33.7537	0.7957	0.3796	0.7446	1.4352	−1.8475	0.0215	1.8319	874.7801	0.298	0.059	3.2978
1.1045	33.2844	1.7732	0.4773	0.6197	1.379	−0.0019	0.0399	1.7732	865.9824	0.377	1.876	2.3804
0.0606	25.4545	1.3783	0.4378	0.6549	1.395	0.0410	0.0404	1.7458	861.8768	0.300	1.478	3.0935
1.0538	32.9032	1.2102	0.4210	0.0176	1.1079	−0.1622	0.0384	1.9453	891.7889	0.351	1.537	2.9135

In the three considered examples, a set of non-dominated solutions are obtained using NSGA-II for the selected advance machining processes. The results obtained using NSGA-II-based GUI are compared with the results of the previous researchers. In the case of ECM process, a Pareto-optimal set of 24 solutions are obtained as given in Table 9.1 and these solutions are compared with the results of Bhattacharyya and Sorkhel (1998) and Mukherjee and Chakroborty (2013). The solutions obtained by Mukherjee and Chakroborty (2013) for the performance parameter {(MRR, overcut)} were {(1.3230, 0.2290), (0.8186, 0.1896), (1.4489, 0.2346)} for different cases with different weights to the considered objectives. In the case of EMM and ECT processes, two and four solutions are obtained as given in Tables 9.2 and 9.3, respectively. The results of the EMM process are found non-comparable directly with the results of the previous researchers, while the results of the ECT process are found comparable and are better than the previous researchers. In the case of ECT process, using weighted-based optimization approach, El-Taweel and Gouda (2011) and Mukherjee and Chakroborty (2013) obtained the results for {MRR, Ra, and RE} as {0.2980 g/min, 0.3545 g/min, {1.1294 µm, 0.8303µm}, and {5.5432 µm, 2.0456 µm}, respectively. The solutions obtained in all the cases are not biased toward the bound values of the process parameters of the considered processes.

9.4 CONCLUSIONS

A non-dominated set of solutions were obtained using NSGA-II for the selected NTM processes such as ECM, EMM, and ECT processes. The results obtained using NSGA-II-based GUI were compared with the results of the previous researchers. In the case of ECM process, a Pareto-optimal set of 24 solutions were obtained and these solutions are not biased toward the range of the ECM process parameters. Similarly, two and four non-dominated solutions were obtained for the EMM and ECT processes, respectively. The results obtained using the proposed methodology are found effective compared to the previous researchers' results. Furthermore, the developed interface provides alternative solutions for each considered performance characteristic and therefore the user has the flexibility to use these solutions as per the requirements to enhance the desired output. In future, this work can be extended to other continuous-domain problems attempted by the researchers in the past to provide alternative solutions. Thus, NSGA-II-based GUI is successfully applied to the considered examples of advanced machining processes. The production engineer can use this developed GUI to obtain the optimal combination of several process parameters to enhance the production capability of the industry without bothering about programming skills.

REFERENCES

Acharya, B.R., Mohanty, C.P., and Mahapatra, S.S. 2013. Multi-objective optimization of electrochemical machining of hardened steel using NSGA II. *Procedia Engineering.* 51: 554–560.
Asokan, P., Ravi Kumar, R., Jeyapaul, R., and Santhi, M. 2008. Development of multi-objective optimization models for electrochemical machining process. *International Journal of Advance Manufacturing Technology.* 39(1–2): 55–63.

Ayyappan, S., and Sivakumar, K. 2015. Enhancing the performance of electrochemical machining of 20MnCr5 alloy steel and optimization of process parameters by PSO-DF optimizer. *International Journal of Advance Manufacturing Technology*. 82: 2053–2064.

Bhattacharyya, B., and Sorkhel, S.K. 1998. Investigation for controlled electrochemical machining through response surface methodology-based approach. *Journal of Material Processing Technology*. 86: 200–207.

Chapman, S.J. 2008. *Matab® Programming for Engineers*. Thomson Asia Ltd, Singapore.

Deb, K., Pratap, A., Agarwal, S., and Meyarivan, T. 2002. A fast and elitist multiobjective genetic algorithm: NSGA-II. *IEEE Transaction of Evolutionary Computation*. 6: 182–197.

El-Taweel, T.A., and Gouda, S.A. 2011. Performance analysis of wire electrochemical turning process- RSM approach. *International Journal of Advance Manufacturing Technology*. 53: 181–190.

Hahn, B.H. and Valentine, D.T. 2010. *Essential Matlab for Engineers and Scientist*. Elsevier Academic Press, Cambridge, MA.

Jain, N.K., and Jain, V.K. 2007. Optimization of electro-chemical machining process parameters using genetic algorithms. *Machining Science Technology*. 11: 235–258.

Konak, A., Coit, D.W., and Smith, A.E. 2006. Multi-objective optimization using genetic algorithms: A tutorial. *Reliability of Engineering System and Safety*. 91: 992–1007.

Kuriakose, S., and Shunmugam, M.S. 2005. Multi-objective optimization of wire-electro discharge machining process by non-dominated sorting genetic algorithm. *Journal of Material Processing Technology*. 170: 133–141.

Majewski, D.E., Wirtz, M., Lampe, M., and Bardow, A. 2017. Robust multi-objective optimization for sustainable design of distributed energy supply systems. *Computers and Chemical Engineering*. 102: 26–39.

Malapati, M., and Bhattacharyya, B. 2011. Investigation into electrochemical micromachining process during micro-channel generation. *Material Manufacturing Processes*. 26: 1019–1027.

Mukherjee, R., and Chakraborty, S. 2013. Selection of the optimal electrochemical machining process parameters using biogeography-based optimization algorithm. *International Journal of Advance Manufacturing Technology*. 64: 781–791.

Pasupathy, T., Chandrasekharan, R., and Suresh, R.K. 2006. A multi-objective genetic algorithm for scheduling in flow shops to minimize the make span and total flow time of jobs. *International Journal of Advance Manufacturing Technology*. 27: 804–815.

Ramesh, S., Kannan, S., and Baskar, S. 2012. Application of modified NSGA-II algorithm to multi-objective reactive power planning. *Applied Soft Computing*. 12(2): 741–753.

Senthilkumar, C., Ganeshan, G., and Karthikeyan, R. 2011. Parametric optimization of electrochemical machining of Al/15% SiCp composites using NSGA-II. *Transaction of Nonferrous Metals Society*. 21: 2294–2300.

Sohrabpoor, H., Khanghah, S.P., Shahraki, S., and Teimouri, R. 2015. Multi-objective optimization of electrochemical machining process. *International Journal of Advance Manufacturing Technology*. 82: 1683–1692.

Srinivas, N., and Deb, K. 1995. Muilti objective optimization using non-dominated sorting in genetic algorithms. *Evolutionary Computation*. 2: 221–248.

Teimouri, R., and Sohrabpoor, H. 2013. Application of adaptive neuro-fuzzy inference system and cuckoo optimization algorithm for analyzing electro chemical machining process. *Frontier of Mechanical Engineering*. 8(4): 429–442.

Wang, W., Radu, Z., and Hugues, R. 2005. Applying multi-objective genetic algorithms in green building design optimization. *Building and Environment*. 40(11): 1512–1525.

Yang, M. D., Chen, Y. P., Lin, Y.H., Ho, Y.F., and Lin, J.Y. 2016. Multiobjective optimization using nondominated sorting genetic algorithm-II for allocation of energy conservation and renewable energy facilities in a campus. *Energy and Buildings*. 122: 120–130.

Yusoff, Y., Ngadiman, M.S., and Zain, A.M. 2011. Overview of NSGA-II for optimizing machining process parameters. *Procedia Engineering*. 15: 3978–3983.

10 ANN Modeling of Surface Roughness and Thrust Force During Drilling of SiC Filler-Incorporated Glass/Epoxy Composites

Ajith G. Joshi
Canara Engineering College

M. Manjaiah
National Institute of Technology Warangal

R. Suresh
M.S. Ramaiah University of Applied Sciences

Mahesh B. Davangeri
Sahyadri College of Engineering & Management

CONTENTS

10.1   INTRODUCTION

Glass fiber-reinforced polymer (GFRP) composites are finding major applications in several engineering fields due to their superior and tailorable properties at a reasonable cost. A filler is incorporated into GFRP composites to improve the mechanical and tribological properties [1]. The use of GFRP in structural applications and as components in assembly-based products has led to inevitable necessary secondary operations such as drilling. Though near-net-shape manufacturing processes exist, drilling is an inherent process in the manufacturing line of components used for assembly-based structures and machines [2]. Investigation of drillability characteristics of GFRP determines ample scope in the current scientific era. The major response characteristics considered for studying the drillability of GFRP in the available literature have been thrust force and hole quality. Delamination is the major drilling-induced damage caused in components of GFRP due to higher values of thrust force. As a consequence, the quality of produced hole deteriorates. The hole quality is typically characterized using surface roughness parameters. Thus, the optimization of process parameters such as speed, feed, and material parameters of GFRP such as reinforcement percentage, filler percentage, and material types is necessary to yield a better surface roughness.

In the past few decades, a substantial number of studies have reported on the drilling characterization of GFRP. Angadi et al. [3] proposed response surface methodology (RSM)-based quadratic models to analyze and predict the drilling characteristics of cenosphere-reinforced epoxy composites. The addition of cenosphere decreases thrust force and surface roughness during the drilling of composites. Budan et al. [2] illustrated that fiber reinforcement percentage plays a vital role in enhancing the machinability of GFRP. Subsequently, an increase in fiber percentage in GFRP increased tool wear, delamination factor, and surface roughness. Kumar et al. [4] investigated the drilling characteristics of glass/vinylester composites. They employed ANOVA to study the relatively significant effect of drill geometry, cutting speed, and feed rate on drilling force and surface roughness. Basavarajappa et al. [5] studied the effects of spindle speed and feed on machinability aspects such as thrust force, hole quality, and specific cutting coefficient during the drilling of glass/epoxy composites and SiC-filled glass/epoxy composites. Basavarajappa and Joshi [6] studied the influence of SiC and Gr filler addition on surface roughness of glass/epoxy composites. They reported that SiC and Gr incorporation drastically decreased the surface roughness.

Recent studies have paid attention to soft computing techniques for modeling and optimization of drilling parameters. Based on the review work, Pontes et al. [7] stated that the artificial neural network approach is appropriate and more suitable to acquire knowledge of materials, process, and other parameters involved in machining. Muthukrishnan and Davim [8] attempted to optimize the machining parameters during turning of metal matrix composites through ANN and ANOVA. They concluded that optimization using ANN is more efficient and less time-consuming. Dhawan et al. [9] presented the ANN model to predict thrust force and torque during drilling of GFRP laminates with four different types of carbide tools. The ANN model exhibited a good efficiency in prediction; besides, they stated that a larger dataset yields better predictive models. Palanikumar et al. [10] depicted that a well-trained ANN is capable of predicting surface roughness during the drilling of GFRP. Kant [11] attempted to model and optimize cutting parameters to attain minimum drilling-induced delamination. The data were captured in accordance with Taguchi's orthogonal array, and the model was developed through the ANN approach to correlate drilling parameters with delamination. The model was able to predict delamination, given a parameter combination.

Velumani et al. [12] presented the mathematical modeling and prediction of thrust force and torque in the drilling of sisal/glass/vinylester composites using RSM and various neural network approaches such as MLPNN, RBFN, and ENN. The study depicted that the results predicted based on MLPNN approach exhibited a good agreement with the experimental results. Nagaraja et al. [13] concluded that GA-MLPNN is a better prediction tool compared to RSM for predicting the drilling characteristics of carbon fiber-reinforced polymer composites. Shunmugesh and Pannerselvam [14] presented an optimization model based on ANN coupled with particle swarm optimization and gravitational search algorithm (PSOGSA) and genetic algorithm (GA). The study showed that PSOGSA is a better optimization tool compared to GA. Balaji et al. [15] developed a model to predict the influence of cutting parameters on thrust force, torque, and delamination factor during drilling of zea fiber-reinforced polyester composites. Feito et al. [16] analyzed the drillability of carbon fiber-reinforced polymer composites as a function of tool wear and drilling parameters. The study implemented ANN approach for simulation of tool wear and prediction of thrust force. Furthermore, ANN can be used as a tool for machining optimization. However, the dataset comprised of broad knowledge of input parameter types and their range is necessary to establish a good model. Karnik et al. [17] presented the multi-response optimization of drilling parameters during the drilling of glass/epoxy composites incorporated with SiC filler using the simulated annealing approach. They also stated that SiC-filled glass/epoxy composites exhibited a better machinability compared to unfilled glass/epoxy composites. Anand et al. [18] investigated the drilling parameters on hybrid polymer GFRP nanocomposites. The study was conducted using Taguchi's technique, and analyses were performed with grey relational method, regression, fuzzy logic, and ANN approach. Among the studied techniques, ANN proved to be better and suitable for process modeling. Vasudevan et al. [19] presented a feedforward backpropagation algorithm-based ANN model to predict thrust force and delamination, which showed better prediction results close to actual values. Wang and Jia [20] employed a hybrid method integrating ANN, GA-based technique, and fuzzy-based technique for multi-objective optimization during drilling of CFRP laminates. The ANN model exhibited a satisfactory prediction with greater accuracy for given cutting conditions.

From the literature review, it can be understood that several studies have reported on the drilling characteristics of GFRP composites, but only limited studies have reported on the drillability of fiber-reinforced composites (FRCs) incorporated with a secondary filler material. The state-of-art research of ANN approach-based predictive modeling of FRCs suggests that ANN modeling has yielded better models for the prediction of drillability characteristics for given input parameters. However, studies on predictive modeling of drilling characteristics of filler material-incorporated

FRCs are lacking. Therefore, the present study aims to develop and present the multilayer perceptron (MLP)-based ANN model to predict the thrust force and surface
roughness. The responses thrust force and surface roughness are correlated with cutting parameters speed, feed rate, and percentage of filler material.

10.2 MATERIALS AND EXPERIMENTATION

10.2.1 MATERIALS

In the present study, composite materials prepared using 7-mil E-glass fibers (supplied by Suntech Fibre, India) were reinforced in a medium viscosity epoxy matrix
(LAPOX, L-12) coupled with room temperature curing polyamine hardener (K-6;
supplied by Atul India Ltd., India). The 400-mesh size SiC particles were used as the
filler material. The composite plates were prepared through the hand layup process
of dimension 300 mm × 300 mm with thickness 8 mm. They were cured under ambi-
ent conditions for 24 h as reported elsewhere [6]. The material composition details of
the different specimens prepared are illustrated in Table 10.1.

10.2.2 DRILLING TEST

The dry drilling experiments were carried out in a Mitsubishi-made M-VSC CNC
machining center. The tool materials used for the drilling test were uncoated solid
carbide and TiN-coated solid carbide. An 8-mm drill tool with the point angle of
118° was used to drill the hole in composite plates. In this chapter, spindle speed,
feed rate, and % of SiC filler are considered as process parameters, while thrust force
and surface roughness are considered as response parameters. A Kistler-made piezo-
electric drill tool dynamometer was used to measure the thrust force. The dynamometer was connected to a personal computer using an analog/digital card. Five levels
of spindle speed (500, 750, 1000, 1250, and 1500 rpm) and feed rate (0.1, 0.2, 0.3,
0.4, and 0.5 mm/rev) were used to conduct the experiments. The surface finish was
measured as a function of surface roughness R_a using a Mitutoyo-made SJ201 digital
surface roughness tester, as shown in Figure 10.1. Two measurements were taken
at each trial, and average readings were considered for the study. The full factorial
design of experiments was employed to acquire the experimental data systematically.
Table 10.2 illustrates the plan of experiments and process parameter combinations
for each trial.

TABLE 10.1
Details of Glass/Epoxy Composite Specimens

Specimen No.	Volume (%)	Volume (%)	Volume (%)
1	50	50	—
2	50	45	5
3	50	40	10

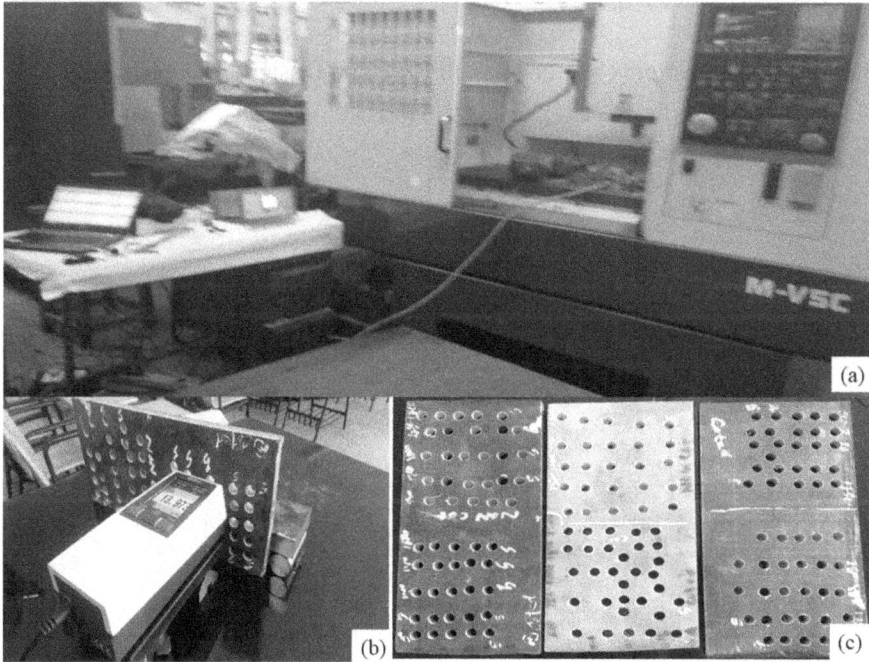

FIGURE 10.1 (a) Vertical machining center used for drilling; (b) surface roughness measurement of the drilled hole; (c) composite samples.

10.3 ANN MODELING AND PREDICTION OF THRUST FORCE AND SURFACE ROUGHNESS

Artificial neural networks are revolutionary computing algorithms that mimic the working of biological brains. It is one of the widely used bio-inspired computing techniques that are efficiently used to model and predict the response of systems. A well-trained, well-established ANN model is capable of solving a wide range of engineering problems, including nonlinear, complex, and multi-dimensional relationships. These networks are generally comprised of fundamental processing elements called neurons [21]. The neurons can store knowledge using experimental data through the learning process and make it available when desired. The knowledge can be utilized to predict the data.

To develop an ANN model to predict the thrust force and surface roughness, a multilayer perceptron neural network (MLPNN) approach was used. A typical MLPNN consists of three layers, namely the input layer, hidden layer, and output layer. The three layers are interconnected by weights. The first layer consists of an input layer with the number of neurons equal to the number of process parameters, each neuron corresponding to each parameter. The second hidden layer consists of a nonlinear function to accomplish a nonlinear mapping. For the training algorithms of MLPNN, there are also many approaches to train MLPNN. The most widely used

TABLE 10.2
Plan of the Experiments as per the Full Factorial Design of Experiments

Trial No.	Speed	Feed Rate	% of SiC	Trial No.	Speed	Feed Rate	% of SiC
1	500	0.1	0		1000	0.3	10
2	500	0.1	5		1250	0.3	0
3	500	0.1	10		1250	0.3	5
4	750	0.1	0		1250	0.3	10
5	750	0.1	5		1500	0.3	0
6	750	0.1	10		1500	0.3	5
7	1000	0.1	0		1500	0.3	10
8	1000	0.1	5		500	0.4	0
9	1000	0.1	10		500	0.4	5
10	1250	0.1	0		500	0.4	10
11	1250	0.1	5		750	0.4	0
12	1250	0.1	10		750	0.4	5
13	1500	0.1	0		750	0.4	10
14	1500	0.1	5		1000	0.4	0
15	1500	0.1	10		1000	0.4	5
16	500	0.2	0		1000	0.4	10
17	500	0.2	5		1250	0.4	0
18	500	0.2	10		1250	0.4	5
19	750	0.2	0		1250	0.4	10
20	750	0.2	5		1500	0.4	0
21	750	0.2	10		1500	0.4	5
22	1000	0.2	0		1500	0.4	10
23	1000	0.2	5		500	0.5	0
24	1000	0.2	10		500	0.5	5
25	1250	0.2	0		500	0.5	10
26	1250	0.2	5		750	0.5	0
27	1250	0.2	10		750	0.5	5
28	1500	0.2	0		750	0.5	10
29	1500	0.2	5		1000	0.5	0
30	1500	0.2	10		1000	0.5	5
31	500	0.3	0		1000	0.5	10
32	500	0.3	5		1250	0.5	0
33	500	0.3	10		1250	0.5	5
34	750	0.3	0		1250	0.5	10
35	750	0.3	5		1500	0.5	0
36	750	0.3	10		1500	0.5	5
37	1000	0.3	0		1500	0.5	10
38	1000	0.3	5				

one is the backpropagation neural network algorithm. However, it is not guaranteed that to find a global minimum of error function because of the gradient descent algorithm often falls into the local minimum area. The convergence rate becomes slow at later iterations. Therefore, improved training algorithms are proposed to avoid the disadvantages of gradient propagation. The third layer consists of an output layer with the number of neurons equal to the number of response parameters. Each neuron in the output layer corresponds to each response parameter. Therefore, training of MLPNN with three neurons in the input layer, ten neurons in the hidden layer, and four neurons in the output layer was chosen and was trained for 3000 epochs. The NN architecture plotted using the NN-SVG online tool is shown in Figure 10.2. The training, validation, and testing of a neural network are accomplished by using experimental data, which are shown in Tables 10.1 and 10.2. The mathematical relation between the performance factors and the response is described in Eq. (10.1).

$$Y = f(X, W); V = \sum_i WiXi \qquad (10.1)$$

where Y is the performing parameter, X is the response to the neural network, W is the weight matrix, f is the model of the process that is being used in the training, V represents the induced local field produced, and x as the input signal and was the synaptic weight.

The model that has been used is MLPNN with a hidden layer. The *"TANSIG"* (tan-sigmoid) transfer function was used to calculate the output at the hidden layer and also the output layer. The output was compared with the experimental output

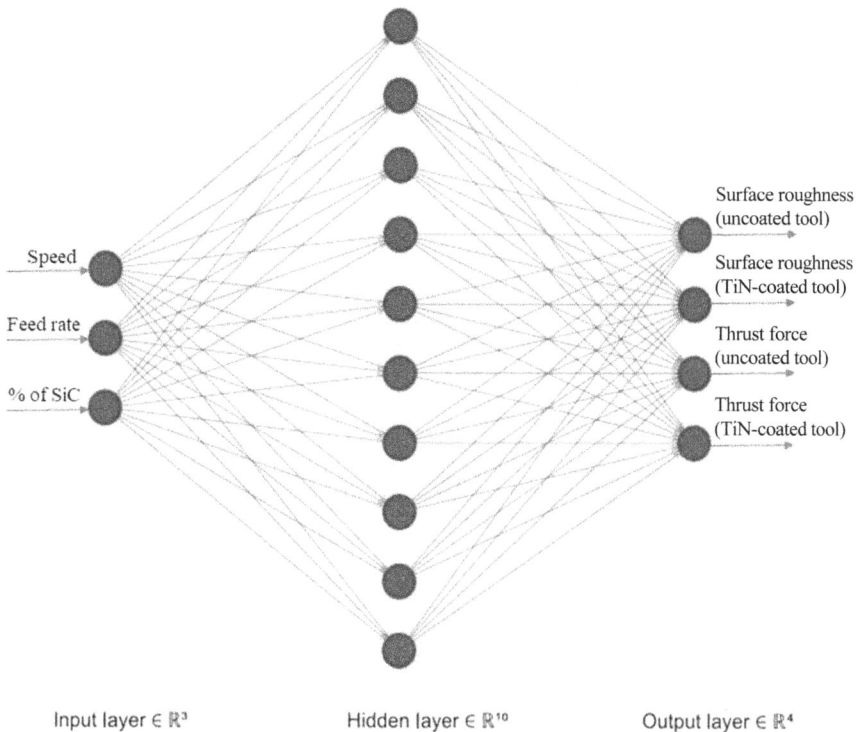

FIGURE 10.2   Diagram illustrating the ANN architecture.

using mean-square error (MSE) as in Eq. (10.2). For training, Levenberg–Marquardt function was used and the threshold was set as 0.0001.

$$\text{MSE, } E = \frac{1}{N} \sum_{0=1}^{N} (T_0 - Y_0)^2 \qquad (10.2)$$

Several networks are constructed, each of them is trained separately, and the best network is selected based on the accuracy of the predictions in the training/validation/testing phases. In the present research work being reported, 53 datasets were considered for training, 11 datasets for validation, and 11 datasets for testing of the developed model. Table 10.3 indicates the trial numbers used for training, validation, and testing of the neural networks. The errors of the validation process are monitored during the training of the network. The test dataset was used for prediction and comparison of the predicted values with the experimental values. Thus, the feasibility of the models was evaluated. The adequacy of the developed ANN models was determined through goodness-of-fit criteria considering the result of the coefficient of determination (R^2) value. It varies from 0 to 1, and the value of R^2 close to 1 indicates the satisfactory adequateness of the developed model.

The detailed procedure as various steps followed for developing ANN models using "*nftool*" available in MATLAB software is presented as below:

Step 1: Initially, datasets required for developing the ANN models were considered, which comprise datasets for training, testing, and validation. Dataset refers to the input dataset and the corresponding target dataset as desired for the training and validation of the ANN model. A large dataset corresponds to the high reliability of the developed ANN model.

Step 2: Decide the number of neurons in the hidden layer. Set the small number of neurons. Ten neurons in the hidden layer were considered in the study.

Step 3: Import the dataset to MATLAB workspace as two variables similar to spreadsheet: One variable comprises of the input dataset and the other variable comprises of the corresponding target dataset.

Step 4: Graphical user interface (GUI) has to be opened by typing "*nftool*" and pressing enter in command prompt window of MATLAB. The input and target data need to be retrieved and chosen.

TABLE 10.3
Dataset Split with the Trial Number Used for ANN Modeling

	Trail Numbers													
Training	1	3	4	5	6	8	10	11	16	17	18	21	22	23
	24	25	26	28	29	30	31	33	35	36	37	38	39	40
	41	43	44	45	46	47	49	50	53	54	55	56	57	58
	60	62	63	64	65	66	67	71	72	73	75			
Validation	9	12	13	14	20	27	59	61	68	69	74			
Testing	2	7	15	19	32	34	42	48	51	52	70			

Step 5: In general, it is always preferred to classify the available large dataset into 70% for training and the remaining for validation and testing. However, in some cases, validation is eliminated and the complete remaining dataset is used for testing. In the current work, considering general thumb rule, 70% of the experimental dataset was used for training the network, while 15% for validation and the remaining 15% dataset for testing the network. The split of the dataset was achieved through the *"dividerand"* command of MATLAB. It randomly divides the complete set of available data for the desired percentage. Hence, select the validation and testing percentage of the dataset in this step.

Step 6: Define a fitting neural network by inputting the number of neurons in the hidden layer. In the current work, it was entered as 10.

Step 7: Later, chose the training algorithm. Three algorithms are available at this stage; for the current study, LM algorithm was selected. Though other algorithms, viz. *"Bayesian Regularization"* and *"Scaled Conjugate Gradient"* were tried, they yielded poor models for training and prediction. The parameters considered for training is illustrated in Table 10.4.

Step 8: Extract the output by saving the results, and compare the obtained output with the desired output dataset. Determine the errors of the output neurons as:

Error = desired output dataset–obtained output dataset.

If the deviation is within the permissible range, then the ANN model is to be considered for prediction; else, improve the model by varying the hidden

---

TABLE 10.4
Neural Network Details Considered in the Study

Network type	Multilayer perceptron (MLP) architecture with feedforward backpropagation network
Network architecture	3–10–4
No. of hidden layers and no. of neurons in the hidden layer	One hidden layer and 10 neurons
Training function used	Levenberg–Marquardt (LM)

Training Parameters	
Maximum number of epochs to train	1000
Performance goal	0
Maximum validation failures	6
Minimum performance gradient	$1e^{-7}$
Initial mu	0.001
Mu decrease factor	0.1
Mu increase factor	10
Maximum mu	$1e^{10}$
Epochs between displays (NaN for no displays)	25
Generate command-line output	False
Show training GUI	True
Maximum time to train in seconds	Inf.

neuron and repeat from Step 6 till the desired goal is achieved. Further, regression analysis is to be carried out to find R^2 value, since it depicts the adequacy of the model for predictions.

Step 9: The developed ANN model could be utilized to predict output for the given input dataset.

10.4 RESULTS AND DISCUSSION

10.4.1 EXPERIMENTAL RESULTS

In the present study, two types of cutting tool materials, viz. uncoated solid carbide tool and TiN-coated carbide tool, were used during the drilling of composites. The obtained experimental results (Table 10.5) revealed that the TiN-coated carbide tool has yielded better machinability characteristics in terms of reduced thrust force and surface roughness. The coating also increased the stability of the carbide tool and enhanced its performance. It might be due to the reason that TiN coating acts as a lubricant, thereby reducing the friction at tool–workpiece interface [22]. Thus, aids for cutting of abrasive fibers as well as plowing and cutting hard filler materials are present in the prepared GFRP specimen. The captured experimental results (Table 10.5) illustrate that an increase in the percentage of reinforcement of filler material (SiC) results in a small increase in the thrust force. Further, the thrust force increased significantly with an increase in the feed rate and decreased relatively with an increase in the speed. The increase in feed rate results in the increase in energy required per unit for the cutting process. The incorporation of filler material imparts hardness to the composites; however, the higher friction generated between the particles and tool causing temperature elevation at workpiece–tool interface facilitates ease of material removal. Hence, a higher thrust force was recorded for a higher level of feed rate and less significant variation was found for the increase in SiC filler reinforcement.

The surface roughness was found to increase significantly with the increase in the feed rate and to decrease with the increase in the surface roughness. The increase in the feed rate leads to a higher pullout force and subsequently to chattering and incomplete machining [6] in GFRP laminates. As a result, fiber pullout and delamination were commonly observed causing a higher surface roughness. The increase in SiC filler percentage has reduced the surface roughness. In case of unfilled GFRP composites, fiber pullout, fiber debonding and fuzzy surface texture were observed [2]. Perhaps, the pullout of fiber and the minimum fuzzy surface texture obtained were low in the filler-incorporated composites. The incorporated filler material facilitates fiber cutting and easy material removal due to the imparted material hardness. Besides, a greater resistance to delamination was offered due to the increased energy absorption capacity of the composites. Hence, less delamination and reduced fiber in the produced holes show better hole quality and the holes possessed minimum surface roughness. Therefore, the incorporation of the optimum percentage of filler in GFRP to enhance mechanical and tribological characteristics was also found to be lucrative to enhance the machinability performance.

TABLE 10.5

Experimental Results for Various Trial Numbers

Trial No.	Uncoated Carbide Tool		TiN-Coated Carbide Tool	
	Thrust Force (N)	Surface Roughness (mm)	Thrust Force (N)	Surface Roughness (mm)
1	33	7.340	22	6.776
2	37	6.942	27	6.127
3	42	6.234	30	5.634
4	20	6.840	14	5.899
5	22	6.123	18	5.776
6	28	5.682	22	5.116
7	16	6.556	12	5.512
8	20	5.823	15	4.899
9	24	4.278	24	4.442
10	12	5.647	10	4.567
11	18	4.655	15	4.199
12	22	4.129	21	3.989
13	13	5.897	14	4.234
14	16	4.340	16	2.890
15	20	3.887	21	2.661
16	48	9.237	33	7.888
17	60	8.128	37	7.389
18	68	7.780	68	6.452
19	34	7.981	25	6.768
20	40	7.436	31	6.211
21	49	6.256	40	5.666
22	23	7.211	21	6.112
23	31	6.349	31	5.434
24	50	4.869	34	4.981
25	28	6.245	19	5.234
26	35	5.342	25	4.556
27	41	4.565	32	4.234
28	21	6.543	18	4.561
29	26	4.889	25	3.445
30	37	4.231	40	2.897
31	71	11.456	44	9.860
32	82	10.780	60	9.320
33	101	9.620	107	8.659
34	50	9.453	42	8.122
35	57	8.983	51	7.657
36	74	7.890	75	6.235
37	45	7.834	30	7.344

(*Continued*)

TABLE 10.5 (*Continued*)
Experimental Results for Various Trial Numbers

Trial No.	Uncoated Carbide Tool		TiN-Coated Carbide Tool	
	Thrust Force (N)	Surface Roughness (mm)	Thrust Force (N)	Surface Roughness (mm)
38	63	6.835	48	6.631
39	79	5.238	65	6.167
40	48	6.997	28	5.889
41	53	5.998	46	5.678
42	61	6.324	48	5.114
43	37	6.779	23	5.871
44	42	5.432	40	4.541
45	49	4.658	48	4.123
46	87	12.368	71	10.879
47	97	11.492	91	10.112
48	136	10.340	130	9.210
49	69	10.342	55	9.235
50	79	9.879	68	8.223
51	110	9.437	96	7.326
52	69	9.328	40	8.112
53	83	8.371	61	7.239
54	103	7.382	84	6.889
55	67	7.673	35	7.912
56	70	6.741	64	7.129
57	78	6.988	75	6.656
58	46	7.436	32	6.341
59	56	6.324	56	5.667
60	55	5.123	60	4.812
61	101	14.675	90	12.453
62	144	13.623	128	11.650
63	152	12.530	154	10.236
64	81	11.230	65	10.223
65	110	10.230	98	8.989
66	147	10.734	114	8.342
67	86	10.327	61	9.768
68	103	9.492	83	8.410
69	138	8.976	113	7.775
70	80	9.324	49	8.675
71	88	8.237	81	7.556
72	96	7.652	90	7.120
73	56	8.498	42	6.881
74	60	7.214	78	5.998
75	65	5.544	72	5.123

10.4.2 REGRESSION ANALYSIS

The experimental data were processed for regression analysis, and quadratic models were established. The regression analysis correlates parameters with the response of the system. Also, it depicts the significant contributing parameters to the response. The established quadratic models are presented as Eqs. (10.3) and (10.4), representing the relationship of the parameters with thrust force and surface roughness during drilling with an uncoated carbide tool, while Eqs. (10.5) and (10.6) represent the relationship of the parameters with thrust force and surface roughness during drilling with a TiN-coated carbide tool. The R^2 and Adj R^2 values of the presented models are greater than 95%, which suggests the good feasibility of models. Further, analysis of variance (ANOVA) was also studied for the developed models with a confidence level of 95%. Thus, the *p-value* of each parameter, a quadratic term of the parameter, or the interaction of parameters less than 0.05 indicates the significance of the corresponding parameter. Tables 10.6–10.9 illustrate the ANOVA for the developed models of thrust force and surface roughness for the uncoated and TiN-coated carbide tools. The ANOVA illustrates that *p-values* for individual parameters, viz. speed, feed rate, and percentage of SiC, and speed vs. feed rate interaction parameter are <0.05, suggesting their significant effect on the response studied. However, the other interaction parameters and the quadratic term of speed are found to be significant in certain conditions; indeed, an insignificant influence was found to exist in majority of the cases.

$$\text{Thrust force} = 1.8033 - 0.004741 \times S + 256.243 \times F + 1.962 \times \text{SiC} - 0.118667 \times S \times F$$

$$- 0.002424 \times S \times \text{SiC} + 7.14 \times F \times \text{SiC} + 6.32381 \times 10^{-6} \times S^2 + 38.0952$$

$$\times F^2 + 0.0656 \times \text{SiC}^2 \quad R^2 = 0.9604; \text{ Adj } R^2 = 0.9549 \quad\quad (10.3)$$

TABLE 10.6
ANOVA for Uncoated Carbide Tool for Thrust Force

Source	Sum of Squares	df	Mean-Square	F-value	p-value	Remarks
Model	84,324.89	9	9369.43	174.99	<0.0001	Significant
Speed	14,860.33	1	14,860.33	277.54	<0.0001	Significant
Feed rate	57,702.43	1	57,702.43	1077.70	<0.0001	Significant
% of SiC	6821.12	1	6821.12	127.40	<0.0001	Significant
$S \times F$	2640.33	1	2640.33	49.31	<0.0001	Significant
$S \times \text{SiC}$	918.09	1	918.09	17.15	0.0001	Insignificant
$F \times \text{SiC}$	1274.49	1	1274.49	23.80	<0.0001	Significant
S^2	32.80	1	32.80	0.6127	0.4366	Insignificant
F^2	30.48	1	30.48	0.5692	0.4533	Insignificant
SiC^2	44.83	1	44.83	0.8372	0.3636	Insignificant
Residual error	3480.25	65	53.54			
Total	87,805.15	74				

TABLE 10.7
ANOVA for Uncoated Carbide Tool for Surface Roughness

Source	Sum of Squares	df	Mean-Square	F-value	p-value	Remarks
Model	410.40	9	45.60	273.41	<0.0001	Significant
Speed	178.17	1	178.17	1068.27	<0.0001	Significant
Feed rate	170.64	1	170.64	1023.13	<0.0001	Significant
% of SiC	36.76	1	36.76	220.39	<0.0001	Significant
$S \times F$	17.60	1	17.60	105.51	<0.0001	Significant
$S \times SiC$	0.4432	1	0.4432	2.66	0.1079	Insignificant
$F \times SiC$	0.0029	1	0.0029	0.0177	0.8946	Insignificant
S^2	6.15	1	6.15	36.86	<0.0001	Significant
F^2	0.5228	1	0.5228	3.13	0.0813	Insignificant
SiC^2	0.1203	1	0.1203	0.7213	0.3988	Insignificant
Residual error	10.84	65	0.1668			
Total	421.24	74				

TABLE 10.8
ANOVA for TiN-Coated Carbide Tool for Thrust Force

Source	Sum of Squares	df	Mean-Square	F-value	p-value	Remarks
Model	73,778.31	9	8197.59	231.60	<0.0001	Significant
Speed	9440.67	1	9440.67	266.72	<0.0001	Significant
Feed rate	45,518.46	1	45,518.46	1286.01	<0.0001	Significant
% of SiC	11,796.48	1	11,796.48	333.28	<0.0001	Significant
$S \times F$	2754.27	1	2754.27	77.82	<0.0001	Significant
$S \times SiC$	650.25	1	650.25	18.37	<0.0001	Significant
$F \times SiC$	2440.36	1	2440.36	68.95	<0.0001	Significant
S^2	998.88	1	998.88	28.22	<0.0001	Significant
F^2	173.72	1	173.72	4.91	0.0302	Insignificant
SiC^2	5.23	1	5.23	0.1477	0.7020	Insignificant
Residual error	2300.68	65	35.40			
Total	76,078.99	74				

$$\text{Surface roughness} = 9.27654 - 0.0066617 \times S + 17.3057 \times F - 0.155466 \times SiC$$
$$- 0.009688 \times S \times F - 5.326 \times 10^{-5} \times S \times SiC + 0.01086 \times F$$
$$\times SiC + 2.738 \times 10^{-6} \times S^2 + 4.98952 \times F^2 + 0.003398 \times SiC^2 \ R^2$$
$$= 0.9743; \text{Adj } R^2 = 0.9707 \tag{10.4}$$

TABLE 10.9
ANOVA for TiN-Coated Carbide Tool for Surface Roughness

Source	Sum of Squares	df	Mean-Square	F-value	p-value	Remarks
Model	317.02	9	35.22	237.25	<0.0001	Significant
Speed	143.04	1	143.04	963.44	<0.0001	Significant
Feed rate	143.81	1	143.81	968.58	<0.0001	Significant
% of SiC	24.86	1	24.86	167.45	<0.0001	Significant
$S \times F$	3.91	1	3.91	26.34	<0.0001	Significant
$S \times$ SiC	0.0140	1	0.0140	0.0944	0.7596	Insignificant
$F \times$ SiC	0.9543	1	0.9543	6.43	0.0136	Insignificant
S^2	0.2790	1	0.2790	1.88	0.1752	Insignificant
F^2	0.0705	1	0.0705	0.4749	0.4932	Insignificant
SiC^2	0.0827	1	0.0827	0.5573	0.4580	Insignificant
Residual error	9.65	65	0.1485			
Total	326.67	74				

$$\text{Thrust force} = 20.4333 - 0.0549638 \times S + 191.429 \times F + 2.372 \times \text{SiC} - 0.1212$$

$$\times S \times F - 0.00204 \times S \times \text{SiC} + 9.88 \times F \times \text{SiC} + 3.48952 \times 10^{-5}$$

$$\times S^2 + 90.9524 \times F^2 - 0.0224 \times \text{SiC}^2 \quad R^2 = 0.9698; \text{ Adj} R^2 = 0.9656$$

$$(10.5)$$

$$\text{Surface roughness} = 7.3962 - 0.0037498 \times S + 14.2355 \times F - 0.1201 \times \text{SiC}$$

$$- 0.004567 \times S \times F + 9.472 \times 10^{-6} \times S \times \text{SiC} - 0.19538$$

$$\times F \times \text{SiC} + 5.83162 \times 10^{-7} \times S^2 + 1.83238 \times F^2 + 0.0028184$$

$$\times \text{SiC}^2 \quad R^2 = 0.9705; \text{ Adj } R^2 = 0.9664 \quad (10.6)$$

10.4.3 ANN MODELING AND PREDICTION

Using the ANN approach, modeling was carried out for thrust force and surface roughness during drilling of the studied polymer composites with two different cutting tool materials. The number of neurons in the hidden layer was varied to attain a minimum error. The number of neurons equal to ten yielded better results. The correlation of training, validation, and testing is illustrated in Figure 10.3; also, a correlation for the overall dataset is indicated. The coefficient of determination (R^2) values obtained for training, validation, testing, and the overall dataset were 0.9961, 0.9916, 0.9963, and 0.9955, respectively. The average errors for training patterns during ANN performance obtained were 4.27 and 6.45 for surface roughness and thrust force for the uncoated solid carbide tool, while 6.74 and 5.02 for surface roughness

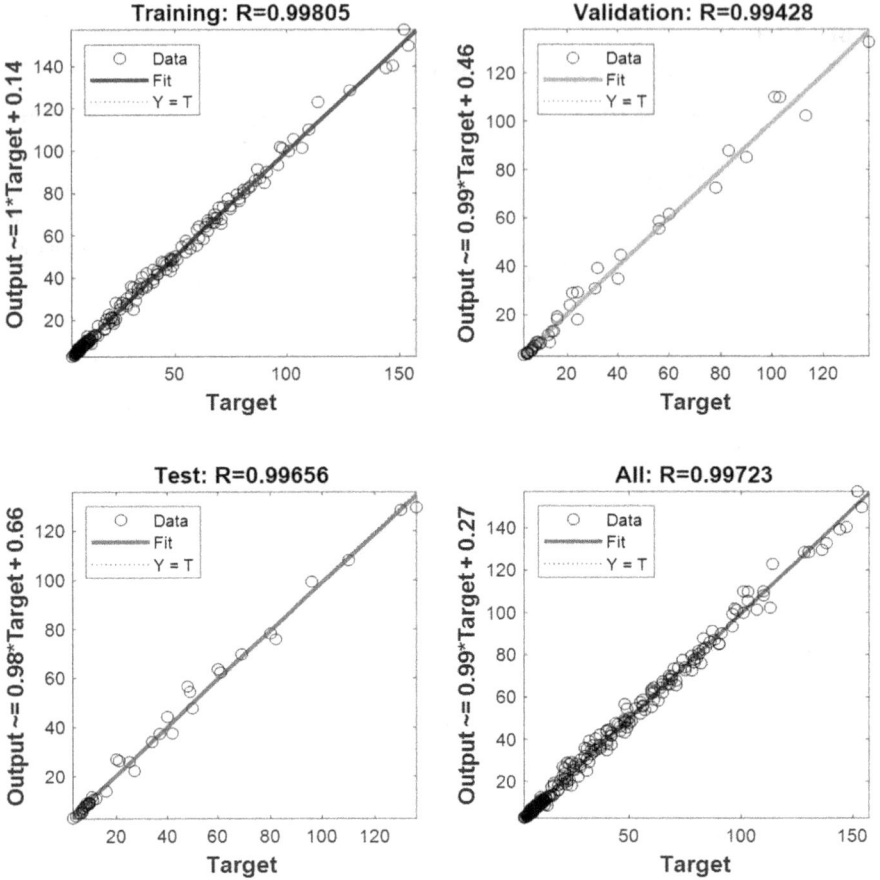

FIGURE 10.3 Correlation of the training, validation, test, and overall patterns.

and thrust force for the TiN-coated solid carbide tool, respectively. The average errors were 7.93 and 5.61 for surface roughness and thrust force for the uncoated solid carbide tool, while 6.94 and 7.04 for surface roughness and thrust force for the TiN-coated solid carbide tool during validation. The average errors were 4.37 and 5.56 for surface roughness and thrust force for the uncoated solid carbide tool, while 6.71 and 5.15 for surface roughness and thrust force for the TiN-coated solid carbide tool during testing of the developed ANN model.

The comparison of ANN results for training, validation, and testing datasets using the MLPNN algorithm is shown in Figures 10.4–10.9. The plots indicate the error is within the acceptable range and a reasonable one. Thus, it can be inferred that the developed ANN model predicted satisfactorily for surface roughness and thrust force during drilling of glass fiber-reinforced polymer composites without and with incorporated filler material. Thus, ANN can be used for prediction, and the network can possibly be generalized sufficiently by better capturing the nonlinear

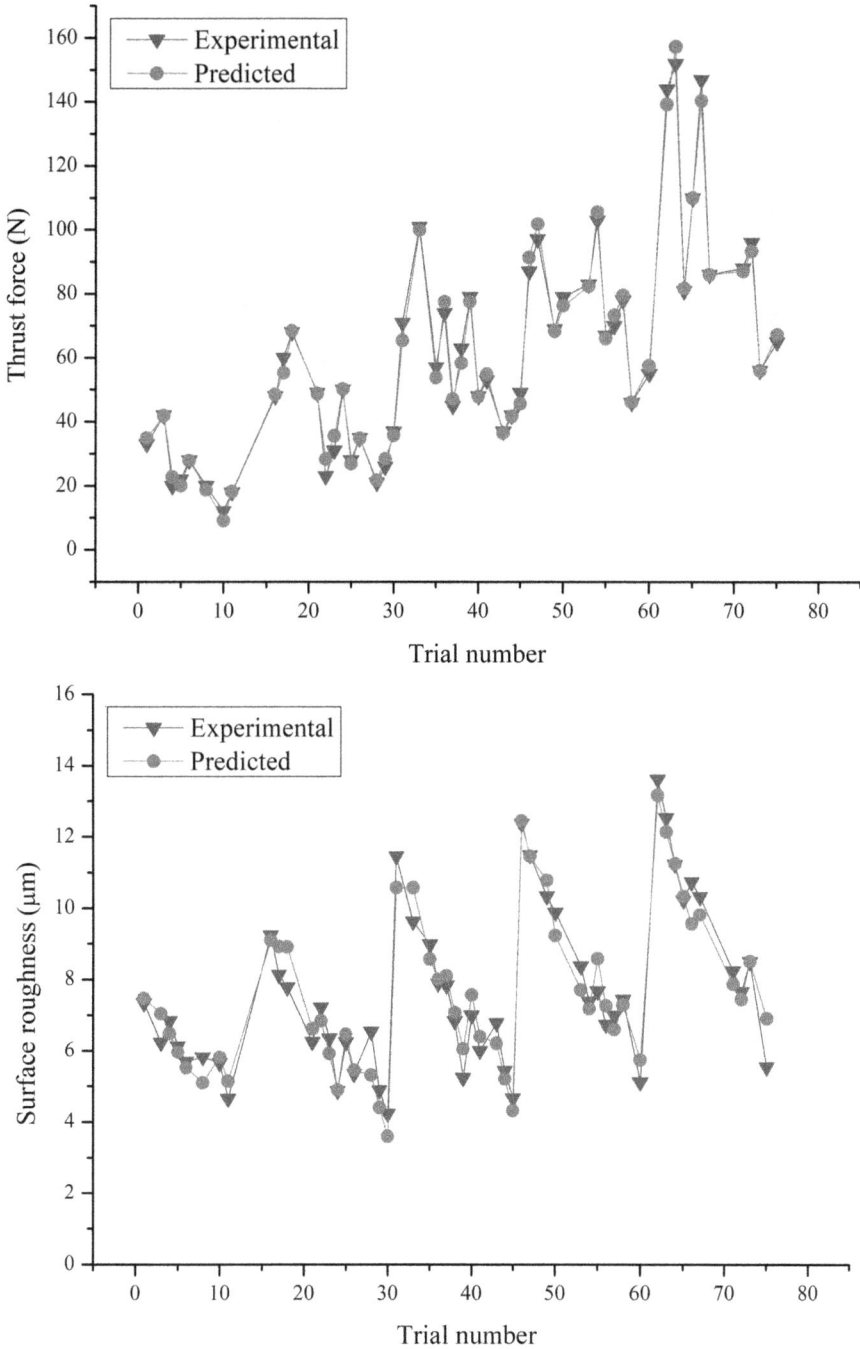

FIGURE 10.4 Comparison of the experimental and ANN outputs for the training set for the uncoated tool.

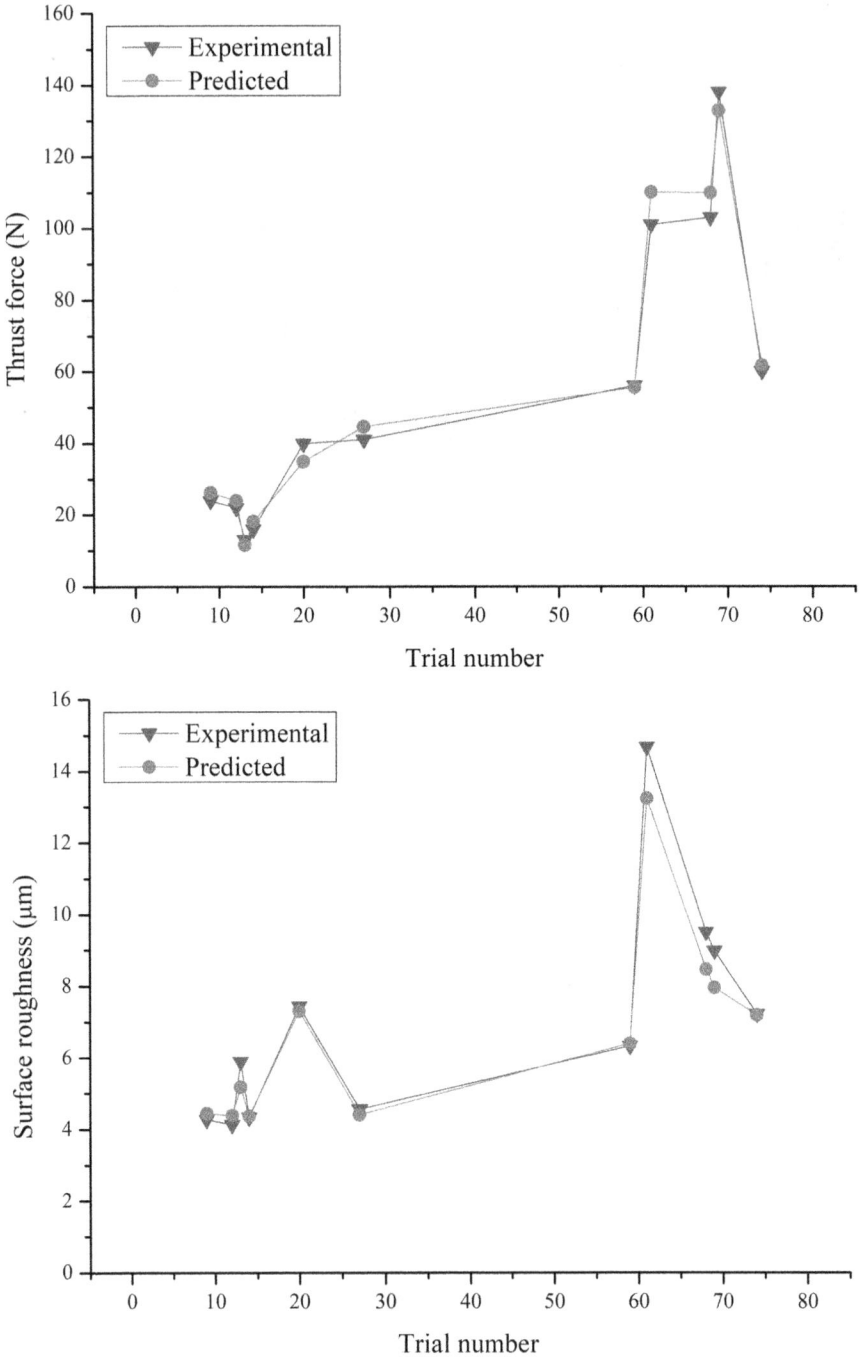

FIGURE 10.5   Comparison of the experimental and ANN outputs for the validation set for the uncoated tool.

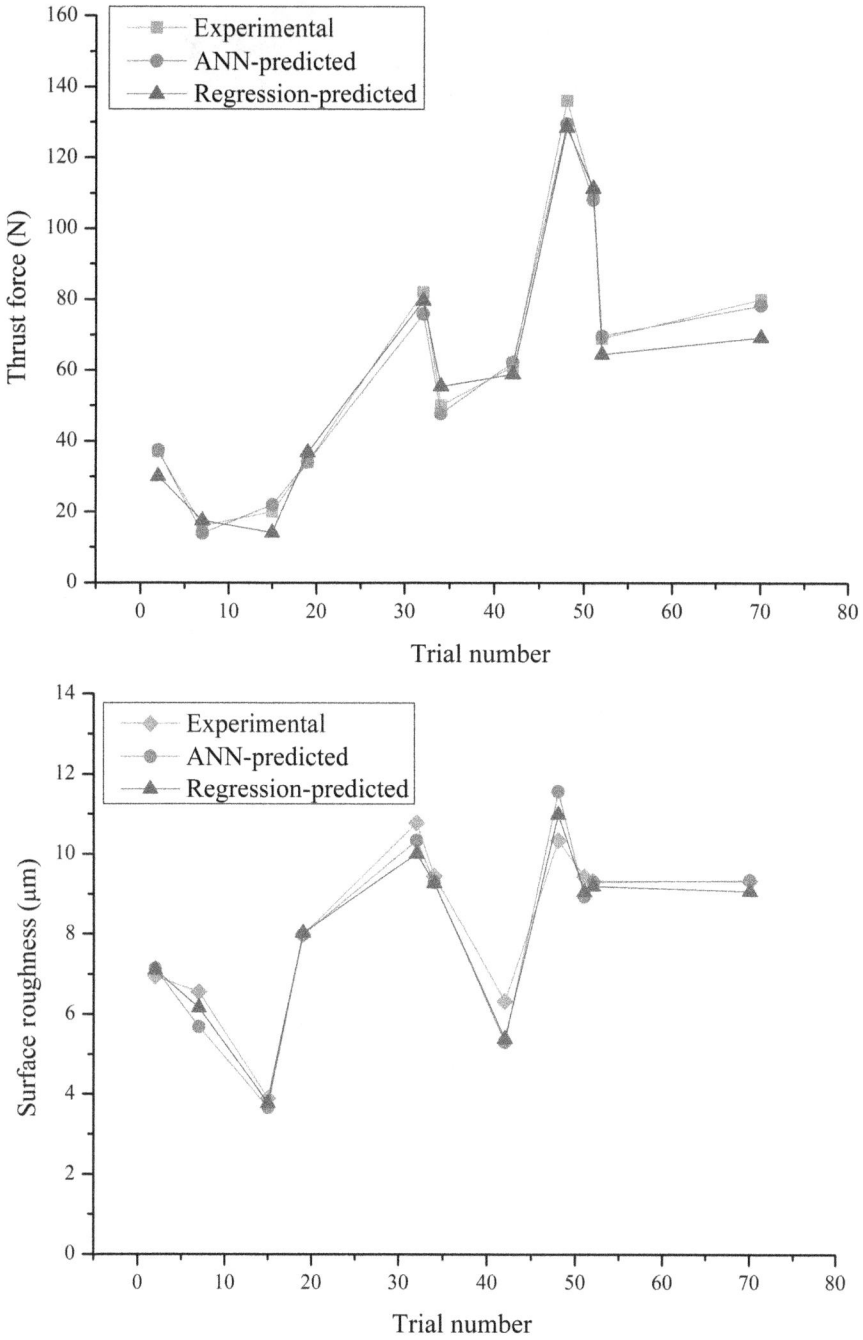

FIGURE 10.6 Comparison of the experimental, ANN, and regression outputs of the testing set for the uncoated tool.

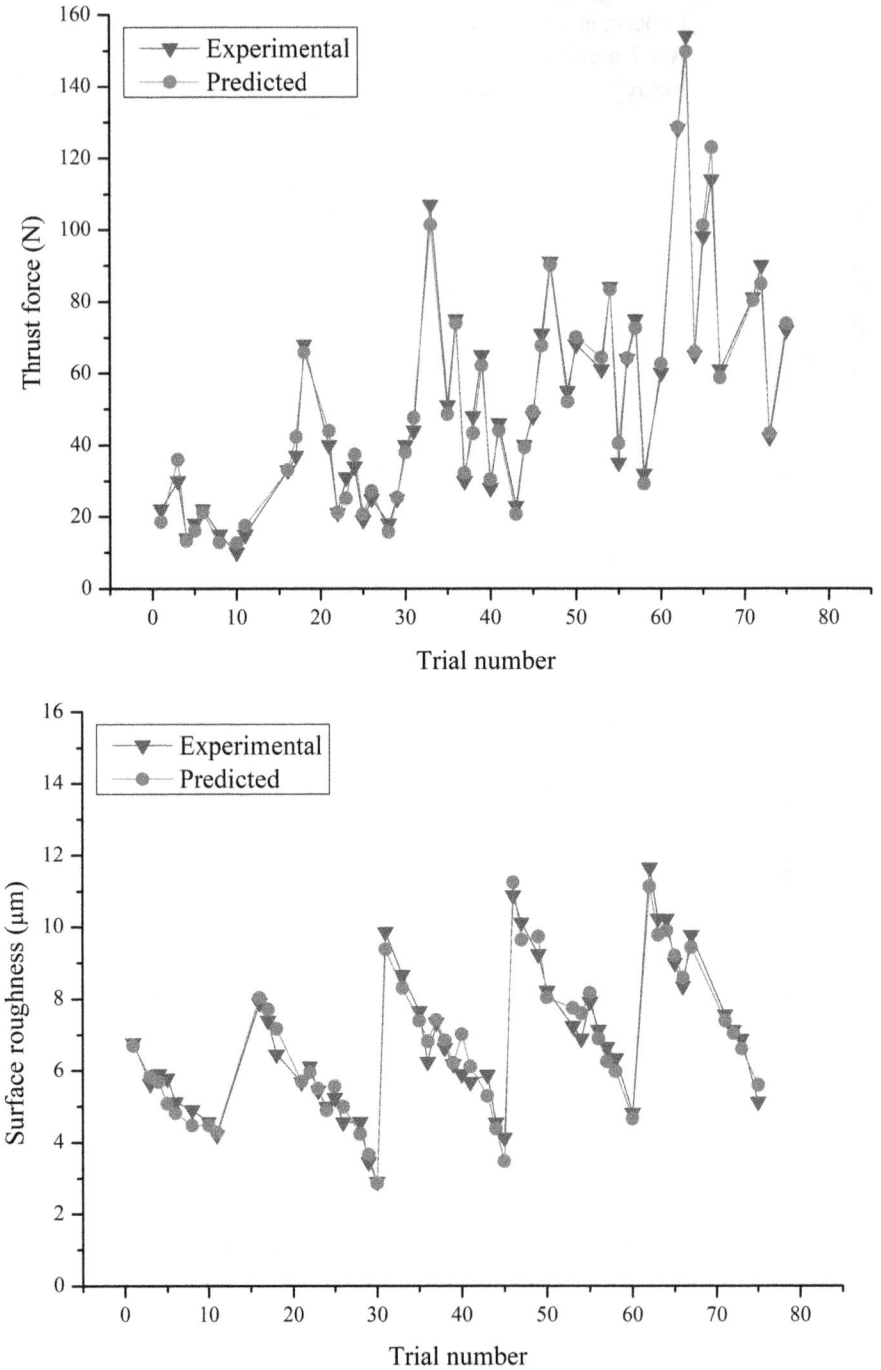

FIGURE 10.7   Comparison of the experimental and ANN outputs for the training set for the TiN-coated tool.

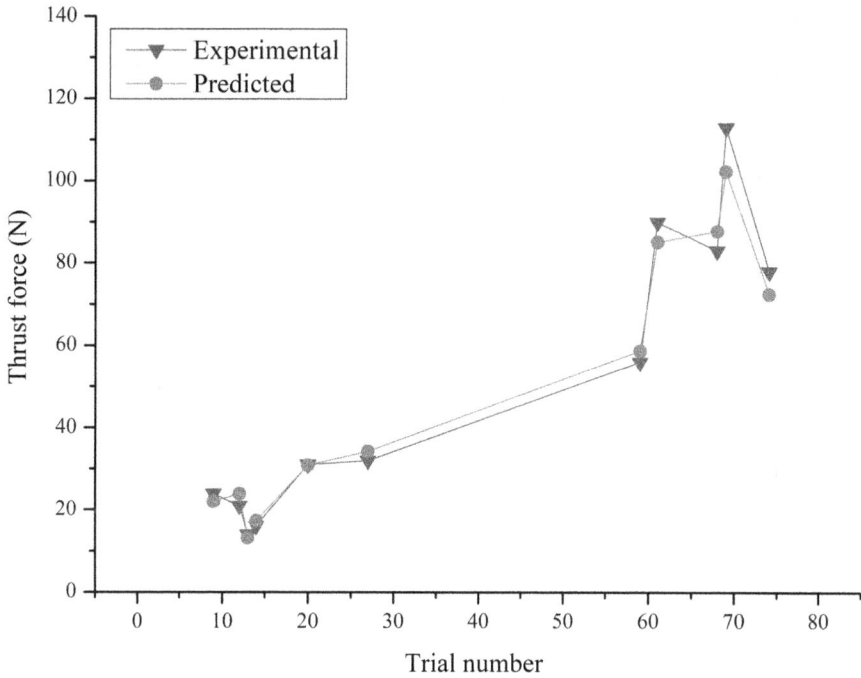

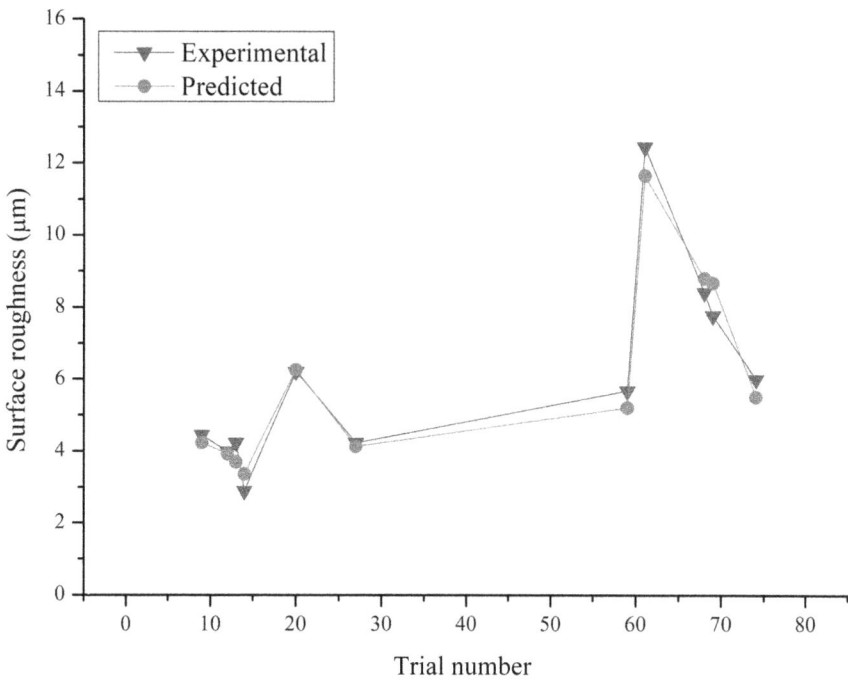

FIGURE 10.8 Comparison of the experimental and ANN outputs for the validation set for the TiN-coated tool.

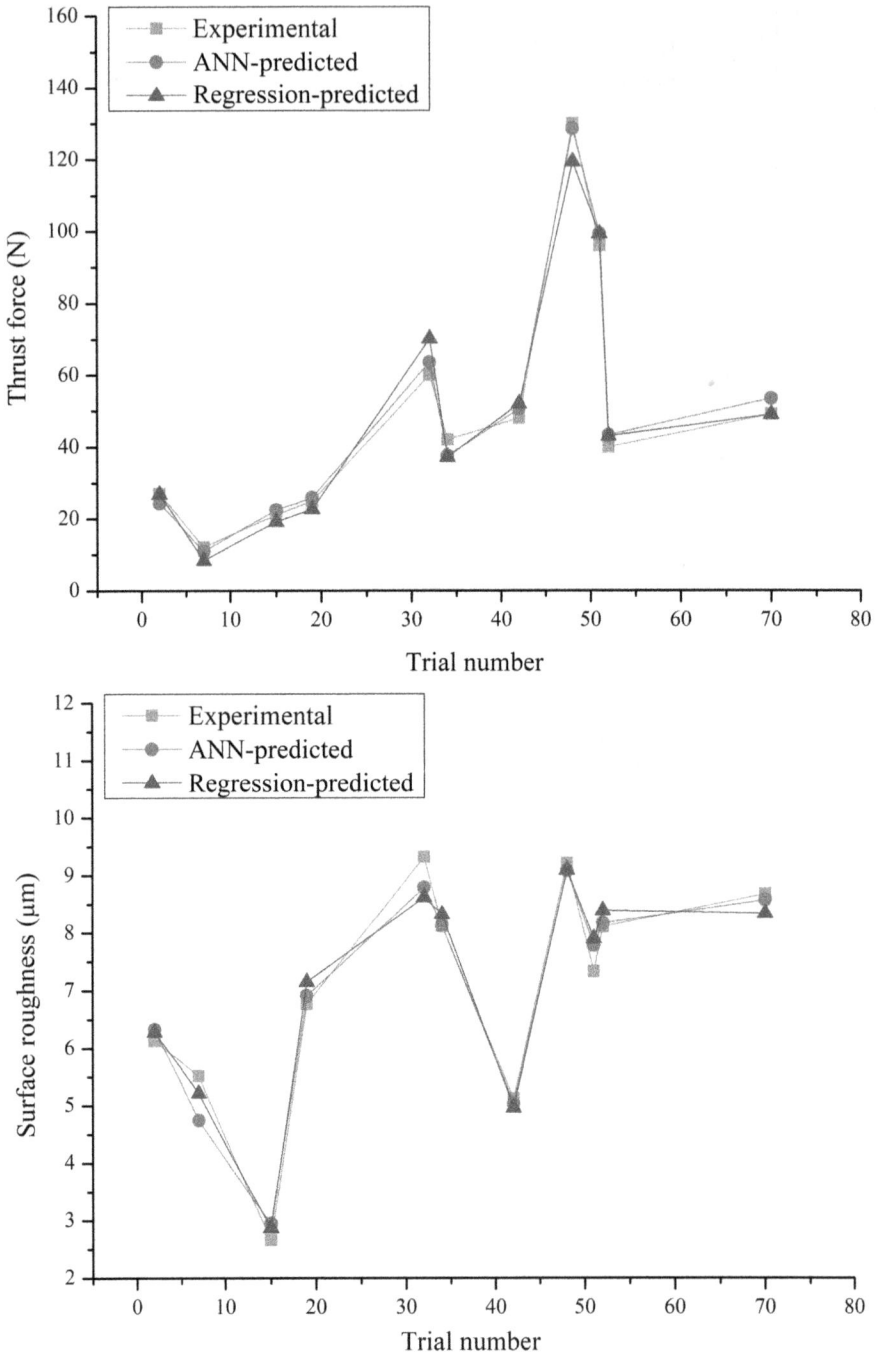

FIGURE 10.9   Comparison of the experimental, regression, and ANN outputs for the testing set for the TiN-coated tool.

relationship of the process parameters with the performance evaluation. Furthermore, Figures 10.6 and 10.9 depict the comparison of values from experimental, ANN prediction, and regression prediction. The plot indicates that the proximity of the ANN-predicted values to experimental values is relatively better compared to the regression-predicted values. Thus, it can be concluded that ANN modeling yields a better model to correlate parameters with the response, indeed also for the prediction of the studied responses compared to quadratic regression models. Subsequently, one ANN model can replace multiple regression equations for correlation and satisfactorily predict multiple responses. Hence, the time consumption for computation and prediction of responses reduces.

10.5 CONCLUSIONS

In this chapter, spindle speed, feed rate, and percentage of SiC reinforcement were considered as the process parameters and thrust force and surface roughness were studied as the response parameters during drilling of glass fiber/epoxy composites with SiC filler. The experimental results were analyzed and predicted using regression models and multilayer perceptron neural network models for pertinent drilling parameters. The major conclusions drawn as per the analysis are as follows:

1. The thrust force was found to significantly increase with an increase in feed rate and decrease less significantly with an increase in spindle speed and percentage of filler reinforcement. An increase in feed rate results in an increase in the energy required per unit, and the incorporation of filler material imparts hardness to the composites. Further, the higher friction generated between the particles and the tool causes a rise in the temperature at the workpiece–tool interface, facilitating easy material removal.
2. Surface roughness increases with an increase in feed rate and decreases with an increase in spindle speed because TiN coating acts as a lubricant during drilling, thereby reducing the friction at the tool–workpiece interface. Also, the incorporated filler material facilitates fiber cutting and easy material removal due to the imparted material hardness. Besides, s greater resistance to delamination was offered due to the increased energy absorption capacity of the composites. Hence, the incorporation of filler material was lucrative in improving the hole quality and a minimum surface roughness was observed for a higher percentage of SiC incorporation (10%).
3. Regression analysis was carried out, and quadratic regression equations were presented to correlate the parameters with the responses studied during drilling of both the types of tools used in the study. The R^2 of 96% for thrust force and 97% for surface roughness for both the types of tools indicates the feasibility of the models.
4. ANOVA revealed that the studied individual parameters, viz. speed, feed, and % of SiC filler, and speed vs. feed rate interaction parameter have a statistical and physical significance on the surface roughness and thrust force.
5. Based on the *p-value (<0.05)*, feed was the most significant parameter for thrust force for both the coated and uncoated tools, while speed was the

most significant process parameter for surface roughness during drilling using the coated and uncoated tools.

6. The developed MLP neural network was appropriately trained and has an average error of 4.37% and 5.56% for surface roughness and thrust force for the uncoated carbide tool. And for the TiN-coated tool, the average error is 6.71% and 5.15% during drilling of glass/epoxy composites. Further, a comparison of the experimental values, ANN-predicted values, and regression-predicted values was carried out. The values predicted using the developed ANN model are relatively in close proximity to the experimental results compared to the values of regression prediction. Hence, the developed MLP neural network model can satisfactorily be used for the prediction of output within a constrained domain.

REFERENCES

1. S. Basavarajappa, A.G. Joshi, K.V. Arun, A. P. Kumar, M.P. Kumar, Three-body abrasive wear behaviour of polymer matrix composites filled with SiC particles, *Polym. Plast. Technol. Eng.* 49 (2009) 8–12. doi:10.1080/03602550903206407.

2. D. Abdul Budan, S. Basavarajappa, M. Prasanna Kumar, A.G. Joshi, Influence of fibre volume reinforcement in drilling GFRP laminates, *J. Eng. Sci. Technol.* 6 (2011) 733–744.

3. S.B. Angadi, R. Melinamani, V.N. Gaitonde, M. Doddamani, S.R. Karnik, Experimental investigations on drilling characteristics of cenosphere reinforced epoxy composites, *Appl. Mech. Mater.* 766–767 (2015) 801–811. doi:10.4028/www.scientific.net/amm.766-767.801.

4. S. Kumar, S.R. Chauhan, P.K. Rakesh, I. Singh, J.P. Davim, Drilling of glass fiber/vinyl ester composites with fillers, *Mater. Manuf. Process.* 27 (2012) 314–319. doi:10.1080/10 426914.2011.585489.

5. S. Basavarajappa, A. Venkatesh, V.N. Gaitonde, S.R. Karnik, Experimental investigations on some aspects of machinability in drilling of glass epoxy polymer composites, 25 (n.d.). doi:10.1177/0892705711408166.

6. S. Basavarajappa, A.G. Joshi, Surface roughness on drilling of glass/epoxy composites with SiC/Gr fillers, *J. Mach. Form. Technol.* 6 (2014) 65–77.

7. F.J. Pontes, J.R. Ferreira, M.B. Silva, A.P. Paiva, P.P. Balestrassi, Artificial neural networks for machining processes surface roughness modeling, *Int. J. Adv. Manuf. Technol.* 49 (2009) 879–902. doi:10.1007/s00170-009-2456-2.

8. N. Muthukrishnan, J.P. Davim, Optimization of machining parameters of Al / SiC-MMC with ANOVA and ANN analysis, *J. Mater. Process. Technol.* 9 (2008) 225–232. doi:10.1016/j.jmatprotec.2008.01.041.

9. V. Dhawan, K. Debnath, I. Singh, S. Singh, Prediction of forces during drilling of composite laminates using artificial neural network: A new approach (2015). doi:10.5937/fmet1601036D.

10. K. Palanikumar, B. Latha, V.S. Senthilkumar, J.P. Davim, Application of artificial neural network for the prediction of surface roughness in drilling GFRP composites, *Mater. Sci. Forum* 766 (2013) 21–36. doi:10.4028/www.scientific.net/MSF.766.21.

11. S. Kant, Application of Taguchi OA array and artificial neural network for optimizing and modeling of drilling cutting parameters, *Int. J. Theor. Appl. Mech.* 12 (2017) 1–12.

12. S. Velumani, P. Navaneethakrishnan, S. Jayabal, D.S.R. Smart, Mathematical modeling and prediction of the thrust force and torque in drilling of sisal / glass-vinyl ester hybrid composite using the RSM, MLPNN, RBFN and ENN methods, *Indian J. Eng. Mater. Sci.* 20 (2013) 289–298.

13. S. Nagaraja, M.A. Herbert, R. Shetty, D.S. Shetty, G.S. Vijay, Soft computing techniques during drilling of bi-directional carbon fiber reinforced composite, *Appl. Soft Comput. J.* 41 (2016) 466–478. doi:10.1016/j.asoc.2016.01.016.
14. K. Shunmugesh, K. Panneerselvam, Engineering science and technology, an international journal machinability study of carbon fiber reinforced polymer in the longitudinal and transverse direction and optimization of process parameters using PSO – GSA, *Eng. Sci. Technol. Int. J.* 19 (2016) 1552–1563. doi:10.1016/j.jestch.2016.04.012.
15. N.S. Balaji, S. Jayabal, S.K. Sundaram, A neural network based prediction modeling for machinability characteristics of zea fiber-polyester composites, *Trans. Indian Inst. Met.* 69 (2016) 881–889. doi:10.1007/s12666-015-0571-3.
16. N. Feito, A. Munoz-Sanchez, A. Diaz-Alvarez, J.A. Loya, Analysis of the machinability of carbon fiber composite materials in function of tool wear and cutting parameters using the artificial neural network approach, *Materials (Basel).* 12 (2019) 2747.
17. S.R. Karnik, V.N. Gaitonde, S. Basavarajappa, J.P. Davim, Multi-response optimization in drilling of glass epoxy polymer composites using simulated annealing approach, *Mater. Sci. Forum.* 766 (2013) 123–141. doi:10.4028/www.scientific.net/MSF.766.123.
18. G. Anand, N. Alagurmurthi, R. Elansezhian, K. Palanikumar, N. Venkateshwaran, Investigation of drilling parameters on hybrid polymer composites using grey relational analysis, regression, fuzzy logic, and ANN models, *J. Braz. Soc. Mech. Sci. Eng.* 40 (2018) 214. doi:10.1007/s40430-018-1137-1.
19. H. Vasudevan, R. Rajguru, R. Yadav, Predictive modelling of delamination factor and cutting forces in the machining of GFRP composite material using ANN, In: H. Vasudevan, V. Kottur, A. Raina (eds.), *Proceedings of International Conference on Intelligent. Manufacturing Automation* Lect. Notes Mech. Eng., Springer Singapore, 2019: pp. 301–313. doi:10.1007/978-981-13-2490-1.
20. Q. Wang, X. Jia, Multi-objective optimization of CFRP drilling parameters with a hybrid method integrating the ANN, NSGA- II and fuzzy C-means, *Compos. Struct.* 235 (2020) 111803. doi:10.1016/j.compstruct.2019.111803.
21. S. Haykin, *Neural Networks and Learning Machines*, Prentice Hall, New York, 2008.
22. S.M. Shahabaz, N. Shetty, S.D. Shetty, S.S. Sharma, Surface roughness analysis in the drilling of carbon fiber / epoxy composite laminates using hybrid Taguchi-Response experimental design, *Mater. Res. Exp.* 7 (2020): 1–12.

11 Multi-objective Optimization of Laser-Assisted Micro-hole Drilling with Evolutionary Algorithms

Hrudaya Jyoti Biswal and V. Pandu Ranga
Indian Institute of Technology Bhubaneswar

Ankur Gupta
Indian Institute of Technology Jodhpur

CONTENTS

11.1 INTRODUCTION

Manufacturing processes can be divided into two broad categories: traditional/conventional processes, namely turning, milling, drilling, and grinding operations, and advanced manufacturing processes using non-conventional sources of energy, such as laser machining, ultrasonic machining (USM), abrasive jet machining (AJM), and electrochemical machining (ECM). Also, these advanced manufacturing processes are proving advantageous because new-generation materials in the category of ceramics, polymers, and other hybrid composites are being added every day, which have distinct properties and are difficult to machine by any of the

existing traditional processes. The requirements of enhanced accuracy and precision for industrial products have contributed to the development of certain advanced
techniques. Various modern machining processes are becoming widely used in the
industries, such as electric discharge machining (EDM), abrasive jet machining
(AJM), ultrasonic machining (USM), electrochemical machining (ECM), and laser
beam machining (LBM), including various modified versions of these processes
depending on a particular principle suitable for a range of materials. As these processes exploit specific properties of the material, it puts certain limitations on their
use, too. The high performance and accuracy of these methods rely on various process
parameters involved. The process becomes complex when more than one objective
is involved. Such multi-objective problems can be tackled by the use of optimization techniques. Hence, optimization has become an essential and paramount tool
in many applications such as engineering, business activities, and industrial designs,
as well as research applications. Being highly non-linear in nature, most real-world
problems demand sophisticated optimization tools. The search for an optimum value
in a multi-dimensional field with numerous process variables as well as objective
functions can prove to be a difficult mathematical problem. Three categories of techniques can be used to solve such mathematical problems, which are experimental,
analytical, and soft computing. Experimental and analytical methods being costly
and tedious, researchers in the last decades have diverted their attention to advanced
optimization techniques.

Among the optimization techniques, nature-inspired (NI) optimization tools
are gaining wider popularity because of their ability to combine the harmonious
cooperation and competition in nature-based systems to reach amazing solutions.
Nature-inspired algorithms take into account a set of probable candidate solutions
and iteratively simulate for offspring solutions until it espouses the fittest solution
within the acceptable range. Evolution being an iterative optimization process refines
the parameter settings of the algorithm for the survival of best characteristics. These
principles of nature have been utilized to create various numerical methods, among
which population-based techniques take center stage. The outstanding efficiency of
these methods such as genetic algorithms (GA), invasive weed optimization (IWO),
and particle swarm optimization (PSO) comes from the fact that they diversify the
search space and enable the algorithm to select the best solution among many generated solutions. With their better global search abilities, these optimization approaches
can find global optima more quickly through cooperation and competition among the
population of potential solutions. These advantages of nature-inspired optimization
techniques make them front-runners for use in advanced manufacturing processes.
Rosenburg was the first to report the evolutionary search method in the 1960s [1].
Since then, different evolutionary algorithms have been designed to solve multi-
objective problems. There are two ways through which these complex problems
can be solved. The first approach deals with combining multiple objectives into a
single one by assigning weight to each one. In contrast, the second approach generates Pareto-optimal solutions, non-dominated with respect to each other. These
techniques have been applied to optimize various conventional manufacturing processes such as grinding [2], turning [3, 4], drilling [5], green sand mold casting [6],
and some other engineering problems [7]. In the last decade, process parameters for

non-conventional machining methods such as AJM [8], EDM [9], and ECM [10] have been optimized using nature-based evolutionary algorithms. This present chapter explores in detail two evolutionary optimization techniques, namely genetic algorithms (GA) and particle swarm optimization (PSO), for micro-hole fabrication using a laser source.

In the past decade, laser has widely been explored as an energy source for performing various operations such as machining, welding, cladding, and drilling. It has also become a boon for micro-machining, such as laser micro-milling, laser micro-drilling, and micro-grooving. The laser micro-machining is the process of material removal through the process of ablation. It is of particular use for hard-to-machine materials as the laser beam acts as a non-contact tool. The precise control of the laser beam can facilitate the creation of micro-features. Despite the wide range of research in the field of laser, several defects such as spatter, taper limit, and heat-affected zone (HAZ) limit its application in the industry. Also, the performance of the machining process using laser involves a large number of input variables that need to be closely monitored. Some of the parameters that can affect the dimensional accuracy and quality of the product are laser power, frequency of the source in case of pulsed laser, pulse width, feed given to workpiece, air pressure, scanning speed of laser source, etc. Hence, it is desirable to achieve a proper parameter setting to satisfy the conflicting objectives of the process. From the literature survey, it was observed that researchers had used different versions of laser source on a wide range of materials such as titanium aluminide, AISI H13 steel, zirconium oxide, aluminum titanate, aluminum oxide ceramic, aluminum–magnesium alloys, and tungsten–molybdenum high-speed steel. Very limited cases were found which addressed the optimization of parameters for laser micro-machining on flexible substrates such as organic or inorganic polymers. In most of the cases, response surface methodology (RSM) and design of experiments (DOE) were utilized to arrive at the optimized parameter setting [11–16]. Few researchers attempted to apply GA and PSO to laser micro-machining and micro-drilling. However, a clear relationship between the input and output variables was not reached in almost all cases.

11.2 FORMULATION OF THE PROBLEM

An ytterbium-doped fiber laser system with a chiller unit was used for drilling micro-holes on a polyethylene terephthalate (PET) substrate by connecting it to a computer by an RS232 communication line. Power requirement was provided through the graphical user interface (GUI) of the laser setup having a maximum power of 400 W. The frequency and pulse duration were fed into the manual controller of the laser source, while the standoff distance was controlled through the micrometer fitted along the Z-axis. The X–Y movement of the stage could be controlled manually through the GUI or a program. PET substrate, an excellent commercial thermoplastic polymer, is a serious contender for substrates used in flexible displays due to its mechanical flexibility under bending and buckling. The low cost, thermal stability, surface inertness, and excellent moisture resistance make it a substrate of interest in an array of fields. The drilling of micro-holes on this substrate with higher accuracy and lower cost can widen its applications. Hence, a PET substrate of 2 mm thickness

Laser power ⟶
Frequency ⟶ **Micro-hole fabrication** ⟶ Hole diameter
Pulse width ⟶ **using laser system**
Object distance ⟶ ⟶ HAZ thickness

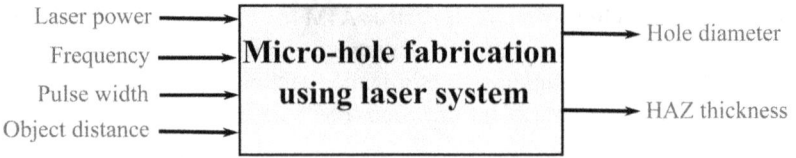

FIGURE 11.1 Input and output variables for the micro-hole fabrication system.

was chosen as the workpiece for the present experiment. From the initial experiments and their analysis, four factors, viz. power of the laser source, frequency, pulse width (duty cycle), and object distance, were considered as input parameters. Figure 11.1 shows the input–output model for micro-hole fabrication using the laser system.

The pulse duration and the frequency were fed into the system through the manual controller of the laser, while the standoff distance was controlled by the micrometer fitted along the Z-axis. An array of holes was produced by regulating the X–Y coordinate through the GUI of the setup manually, which could be programmed, too. The above-mentioned four factors were selected as input variables from the initial experiments. The ranges of values considered for all the input parameters are shown in Table 11.1. A full-scale central composite design (CCD) with 31 experiments was chosen. Five levels were taken for each factor, and five experiments were carried out for each experimental run order. Hence, a total of 135 (31 × 5) micro-hole drilling experiments were performed. Figure 11.2 shows the microscopic images of the initial experiments conducted. Experiments were carried out according to the central

TABLE 11.1

Process Parameters and Their Ranges

Parameters	Unit	Symbol	Range High	Low
Laser power	W	P	36	4
Frequency	KHz	f	5	1
Pulse width	%	W	18	2
Object distance	mm	D	4	0

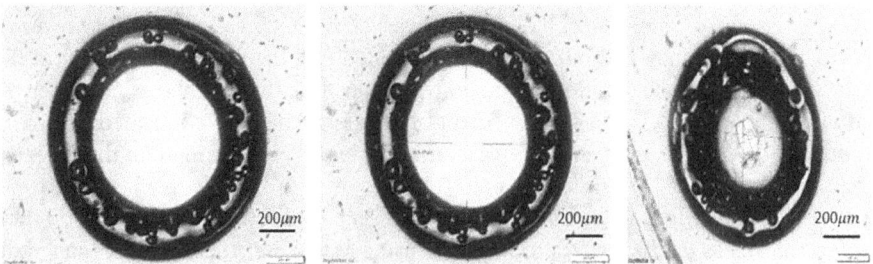

FIGURE 11.2 Microscopic images of the experiments conducted.

composite design, and the hole diameter along with HAZ thickness was measured for each run.

Regression analysis was carried out for both the process variables, and the obtained p-values for all the terms are demonstrated in Tables 11.2 and 11.3. The relationship between the responses and the process variables available in the above-mentioned literature are as given below:

$$\text{Hole Diameter}\,(\mu m) = -203 + 22.8\,P + 200\,f + 28.5\,W - 7.59\,Pf \qquad (11.1)$$

$$\text{HAZ thickness}\,(\mu m) = 54.7 + 9.28\,P + 10.5\,W + 0.274\,DW$$
$$- 0.344\,PW - 0.106\,P^2 + 1.53\,f^2 \qquad (11.2)$$

All linear terms, square terms, and their interaction terms were analyzed for their influence on the hole diameter as well as the HAZ thickness. The regression equations were formulated depending on the p-values, which are seen to be less than 0.05 (corresponding to a confidence level of 95%). Some insignificant terms were included as without those, there was a lack of fit of the model. Hence, the insignificant terms need not be removed from the model. Moreover, the p-value for the overall analysis is within the value of 0.05. The curves showing the calculated value using the models have a close fit with that of the curves determined using experimental values, as shown in Figure 11.3.

TABLE 11.2
The predictors and p-values for Hole diameter

Predictor	Coefficient	SE Coefficient	T	P
Constant	−202.9	419.7	−0.48	0.633
P	22.80	17.31	1.32	0.201
f	200.1	117.7	1.70	0.103
W	28.491	9.559	2.98	0.007
Pf	−7.593	5.369	−1.41	0.171

TABLE 11.3
The Predictors and p-values for HAZ Thickness

Predictor	Coefficient	SE Coefficient	T	P
Constant	54.74	55.18	0.99	0.332
P	9.283	3.358	2.76	0.012
W	10.482	4.394	2.39	0.027
DW	0.2743	0.4349	0.63	0.535
PW	−0.3435	0.1919	−1.79	0.088
P^2	−0.10599	0.06456	−1.64	0.116
f^2	1.5312	0.7920	1.93	0.067

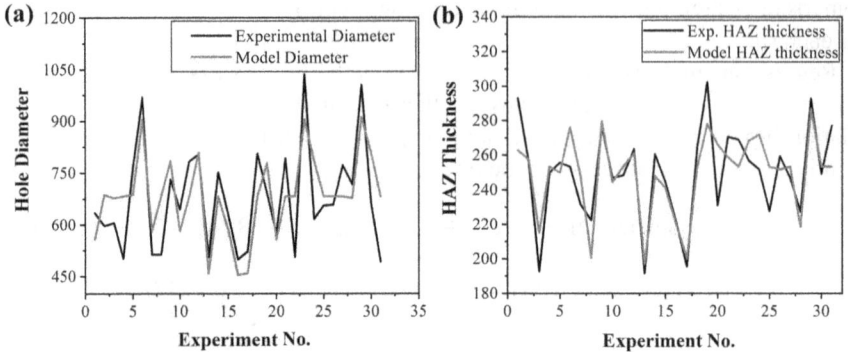

FIGURE 11.3 Graph showing experimental versus calculated (a) hole diameter and (b) HAZ thickness.

An attempt has been made to optimize the process of micro-hole drilling using laser energy with multiple output parameters with the help of evolutionary algorithms. As two variables are involved in the said process, the weighted factor method has been adapted to normalize and combine both to form a single-objective function. Hence, the function to be minimized is expressed as:

$$\text{Minimize } Z = (w_1 \times \text{HD} + w_2 \times \text{HAZ}), \tag{11.3}$$

subject to constraints:

$$4 < P < 36 \tag{11.4}$$

$$1 < f < 5 \tag{11.5}$$

$$2 < W < 18 \tag{11.6}$$

$$0 < D < 4, \tag{11.7}$$

where HD and HAZ are functions for hole diameter and HAZ thickness, while w_1 and w_2 are the weighted factors for HD and HAZ, respectively, and $P, f, W,$ and D are the process variables. The weighted factors are chosen in such a way that their sum is equal to one. A higher weighting factor is assigned to the variable, which is thought to have more importance.

11.3 USE OF NATURE-INSPIRED ALGORITHMS FOR OPTIMIZATION

In the present chapter, nature-inspired evolutionary optimization techniques have been explained and employed to optimize the developed single-objective function (Eq. 11.3). All NI algorithms operate through similar search characteristics and algorithm dynamics.

- All evolutionary optimization techniques use a solution vector in the form of an agent, such as particle, ant, cuckoo, and bat. Certain size of these agents represents a population which accounts for the diversity and different values of fitness of the problem.
- Some operators (such as crossover and mutation) in the form of equations are put to use to achieve the iterative evolution of the population. The convergence of the system is determined by the evolution of the solutions with different properties.
- The moves of the agent are perturbed by randomization techniques such as mutation to escape any local optima, thus minimizing the risk of getting stuck locally.
- Selection and elitism implement the principle of "survival of the fittest" so that the best solution can be carried forward in the population generation after generation. It acts as the driving force to achieve convergence by reducing diversity in a structured way.

11.3.1 GENETIC ALGORITHMS

Genetic algorithms are population-based search and optimization techniques that mimic the evolutionary technique of "survival of the fittest" that was first developed by John Holland [17]. The distinct advantages such as being gradient-free, high exploration capability, and parallelism give GA an edge over other processes. It can deal with various types of objective functions such as linear or non-linear, stationary or non-stationary, and continuous or discontinuous. In GA, binary or decimal numbers are assigned to a point in search space that are known as strings or chromosomes. A set of chromosomes is known as the population. Three fundamental operators, *viz.* reproduction, crossover, and mutation, are utilized, and together, they constitute one generation. The roles of these operators are very different from each other:

- *Crossover*: The crossover is a significant operator between two parents with a probability of crossover, P_c, which is applied by swapping one segment of the parent chromosome with the corresponding segment in another at a random position. Crossover can occur at multiple sites also to increase the evolutionary efficiency of the algorithm.
- *Mutation*: The mutation operator is used to flop the bits at single or multiple sites depending on the probability of mutation, P_m. It is necessary to be implemented in the algorithm to overcome the obstacle of getting stuck in local optima.
- *Selection*: It is based on the fitness of individual solutions so that reproduction can be proportional to fitness. It is used to choose the best solution to be carried to the next generation.

Figure 11.4 shows schematically the steps followed in genetic algorithms. The combined effect of all these operators on the algorithm is complex. Hence, it is important to choose the parameter values judiciously. Crossover restricts the action within a subspace. For example, in the case of a parent string $S_1 = [aabba]$ and $S_2 = [abbab]$,

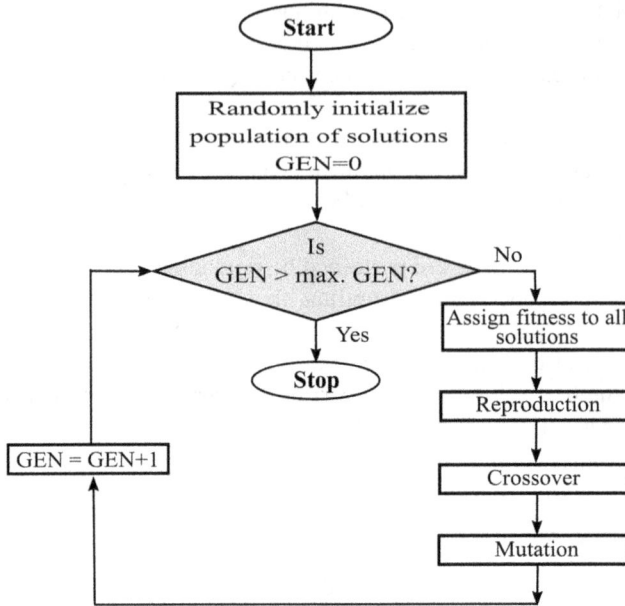

FIGURE 11.4   Flowchart of genetic algorithms.

the offspring string will always be in the form [*a*....] whatever may be the crossover action. However, mutation helps the solution to plunge outside of the subspace. In the previous example, if the *a* of the first-string flips to *b* due to mutation, the solution string will take the form [*b*....], thus increasing the diversity of the population. Crossover or mutation does not make use of the knowledge about the objective function or the fitness landscape. But selection or elitism guides the algorithm through the fitness landscape of the solutions and ensures that the best solution survives. Various selection methods such as roulette wheel selection, tournament selection, linear ranking selection, and random selection are used in genetic algorithms.

The choice of various parameter values is another significant issue for the implementation of the GA. The probability of crossover, P_c, is chosen high, generally in the range of 0.7–1.0. A smaller value for P_c implies that the crossover may happen sparsely, which is not efficient from the point of view of evolution. On the other hand, the mutation probability, P_m, is usually small, in the range of 0.001–0.08. A higher value of mutation makes it difficult for the system to converge. Deciding the right population size (N) is also important as it ensures proper evolution. Too small a size of the population can make it extinct before it has reached the optimal solution. Too large a population can enhance the evaluations of the objective function, thereby extending the computing time.

11.3.2 Particle Swarm Optimization (PSO)

PSO is a widely used metaheuristic optimization technique based on the simulation of the social behavior of birds in a flock developed by Kennedy and Eberhart in 1995 [18]. The current optimization method tracks the pattern that regulates the synchronous flying path of a bird flock and its sudden change with regrouping in an optimal formation. As compared to other population-based techniques, the major advantages of PSO are its simplistic implementation approach requiring less algorithm-specific parameter adjustments, efficiency in terms of computation, and its ability to escape local optima, and overall convergence to the global optimum. Also, swarm intelligence-based algorithms make use of multiple agents as an evolving population, thus showing good adherence to the natural systems that they intend to mimic. In PSO, individuals, called particles, are flown through a search space, which are assigned positions and velocities. The algorithm maintains a swarm of particles that is similar to the population in analogy with evolutionary algorithms. The position of the particle is decided according to its own experience and that of its neighbor. The velocity is dynamically adjusted according to the particle's own knowledge, and that exchanged from the neighborhood or swarm. This experiential knowledge of the particle is indicated as the *cognitive component*, whereas the social knowledge is referred to as the *social component*. The working cycle of the PSO is shown in the schematic diagram in Figure 11.5.

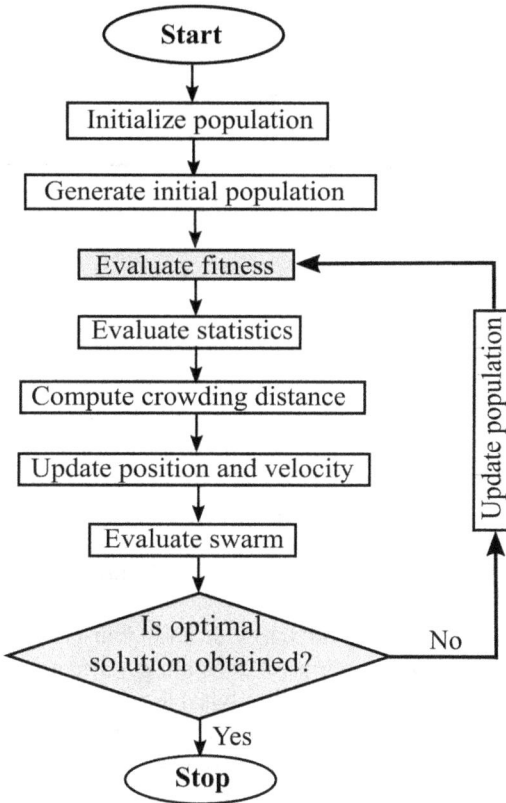

FIGURE 11.5   Flowchart of particle swarm optimization method.

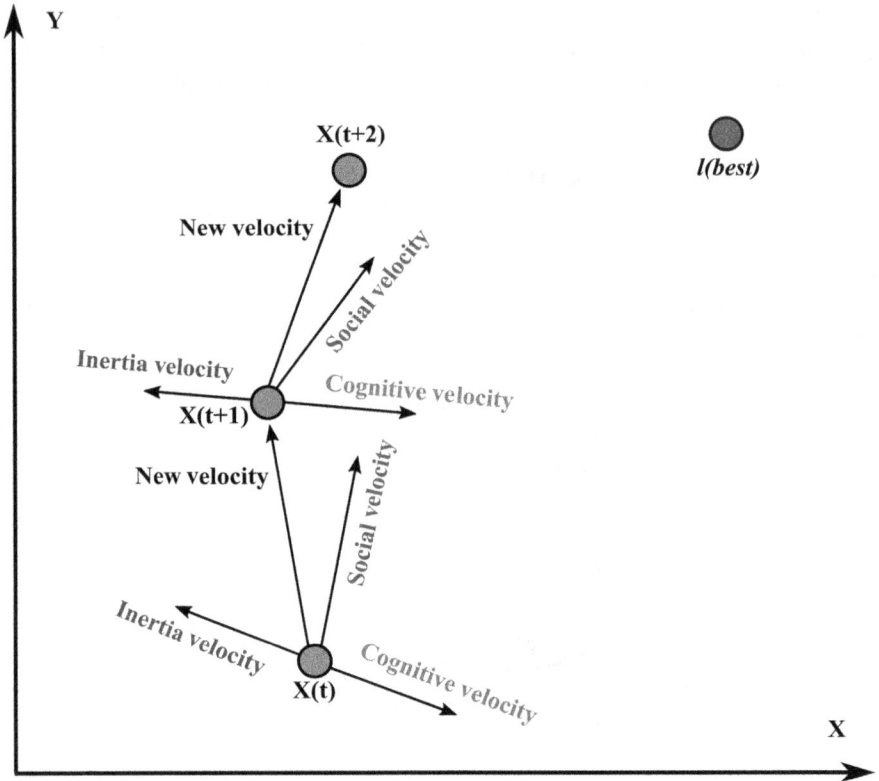

FIGURE 11.6   Illustration of position and velocity update for a single particle.

The particles or agents in a PSO algorithm depend linearly on the position and velocity; hence, in a sense, PSO algorithmic equations are linear in nature. Each particle in a PSO algorithm moves toward its local best, *l(best)*, while simultaneously adjusting its position and velocity toward the global best, *g(best)*; at the same time, it has a tendency for random movement. There is a local best for all the particles, and the objective is to find the global best among the local bests. Figures 11.6 and 11.7 illustrate the movement of particles toward the local best by updating the position and velocity and the subsequent movement of swarms toward the global best.

PSO has the advantage of initializing with a random population since as compared to other evolutionary algorithms, it does not get affected significantly by the initial swarm. But a bigger swarm size is advisable as it speeds up the convergence toward global optima. The particle velocity, too, should be decided judiciously to avoid premature convergence. The inertia weight is chosen carefully to provide enough velocity to the system to move forward.

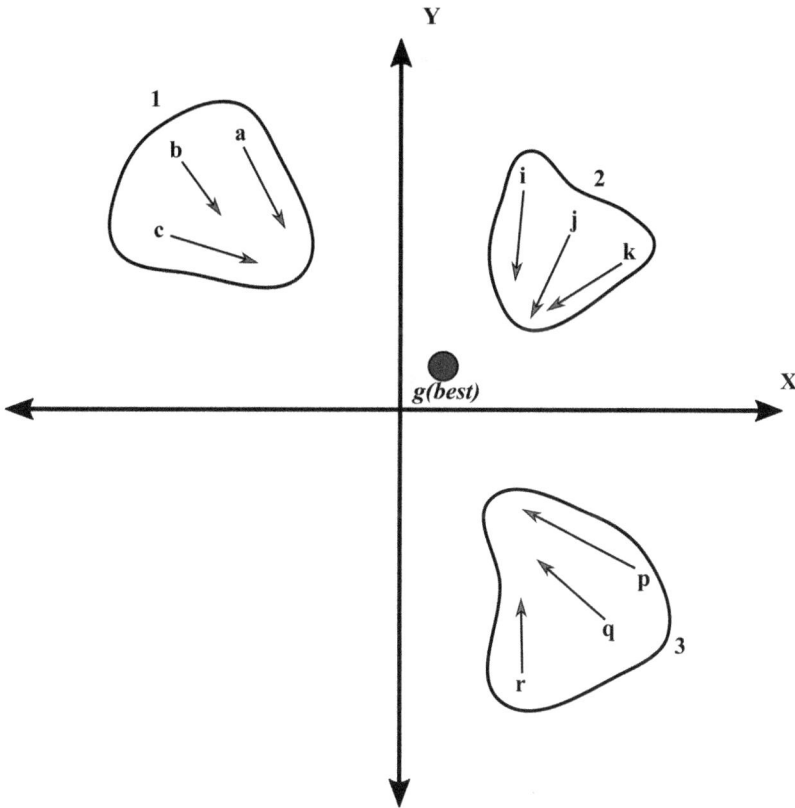

FIGURE 11.7 Schematic representation of swarms moving toward the global best.

11.4 RESULTS AND DISCUSSION

11.4.1 GA APPLIED TO MICRO-HOLE FABRICATION USING LASER ENERGY

A real-coded GA has been used for computer simulations using tournament selection type of reproduction and uniform crossover. The mutation operator was used to arrive at the global minima as the objective of the function is to minimize. To achieve an effective search, there must be a proper balance between GA parameters, namely probability of crossover (P_c), probability of mutation (P_m), population size (N), and maximum number of generations (G). The interaction between them needs to be studied to arrive at their optimal values. As GA parameters are problem dependent, several trials have been attempted to select them in an optimal sense. A parametric study (varying one parameter while keeping the others constant) has been utilized to determine the GA parameters for the optimum value of both the hole diameter and HAZ thickness. The results of the parametric study are plotted in Figure 11.8, and the procedure is explained below.

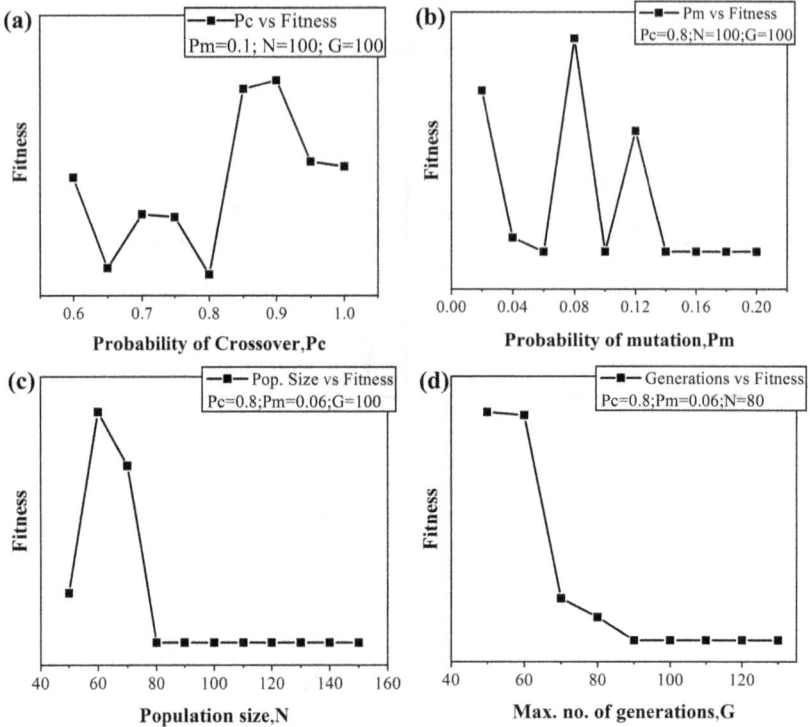

FIGURE 11.8 GA parametric study: (a) P_c vs. fitness, (b) P_m vs. fitness, (c) population size vs. fitness, (d) generations vs. fitness.

The GA parameters were decided to be varied in ranges such as 0.6–1.0 for P_c, 0.02–0.2 for P_m, 50–150 for N, and 50–130 for G. Figure 11.8a shows the variation of fitness values with respect to change in P_c values while keeping P_m, N, and G at a fixed level, generally at the mid-value. As this is a minimization problem, the probability of crossover value (P_c^*) corresponding to the minimum fitness value was chosen. In the next step, the optimum probability of crossover (P_c^*) was considered along with population size and maximum number of generations at the same level. The plot of the probability of mutation (P_m) versus fitness values is demonstrated in Figure 11.8b. The optimum value of P_m (P_m^*) was selected, and both P_c^* and P_m^* were used in the next stage for the determination of optimum population size (pop*), as shown in Figure 11.8c. Finally, the study was conducted to determine the maximum number of iterations (G^*) for which the fitness value is minimum after taking the optimized values of other parameters, $viz.$ P_c^*, P_m^*, and pop*. Figure 11.8d depicts the plot of maximum number of generations versus fitness values. The optimum parameters obtained are mentioned below:

Probability of crossover (P_c^*) = 0.8.
Probability of mutation (P_m) = 0.06.
Population size (pop*) = 80.
Maximum number of generations (G^*) = 90.

TABLE 11.4
Optimum Hole Parameters for Multiple Responses Using GA

Process Parameters and Responses	Unit	Optimum Values
P: power	W	12
f: frequency	Hz	2
W: pulse width	%	6
D: standoff distance	mm	1
HD: hole diameter	μm	459.44
HAZ: HAZ thickness	μm	195.696

Table 11.4 shows the optimum values of parameters to obtain a minimum hole diameter and HAZ thickness for micro-drilling using laser energy. The model has a good convergence for all the weight factors that were tested. As there are two parameters to be optimized, the weight factor does not have a significant effect on the fitness value. Hence, equal weights were given to both the factors.

11.4.2 PSO Applied to Micro-hole Fabrication Using Laser Energy

This chapter puts to use a multi-objective PSO (MOPSO), a variant of PSO, to obtain the optimum process parameters for the micro-hole drilling process. The two characteristics of the particle, i.e., position and velocity, are updated through many generations until the desired solution is achieved. The swarm size selection method is utilized, which helps in reaching the local best and avoids pre-convergence of the problem. Initially, a swarm is generated with position X_i and velocity V_i ($i = 1$, $2...$, P), where i refers to the process control parameters. The objective function is defined for each local best [$f(l$best)] and global best [$f(g$best)]. The velocity is updated based on the previous one taking into account both the cognitive and social components. The new velocity is given by the formulation:

$$V_i^t = WV_i^{t-1} + C_1U_1\left(l\text{best} - x_i^t\right) + C_2U_2\left(g\text{best} - x_i^t\right) \tag{11.8}$$

The new particle position is $x_i^t = x_i^{t-1} + V_i^t$, where x_i^t is the current position of the particle, W is the inertia weight, C_1 and C_2 are the self-adjustment and social adjustment learning factors, respectively, and U_1 and U_2 are random vectors. The parameters, namely swarm size, the number of generations, and inertia weight, play an important role in the present approach. A systematic study was conducted by varying one parameter and keeping others constant. The subsequent results are shown in Figure 11.9. The parameters of the PSO responsible for the optimum performance of the function are as follows:

Inertia weight (W^*) = 1.0.
Swarm size (SS^*) = 70.
Maximum no. of generations (G^*) = 130.

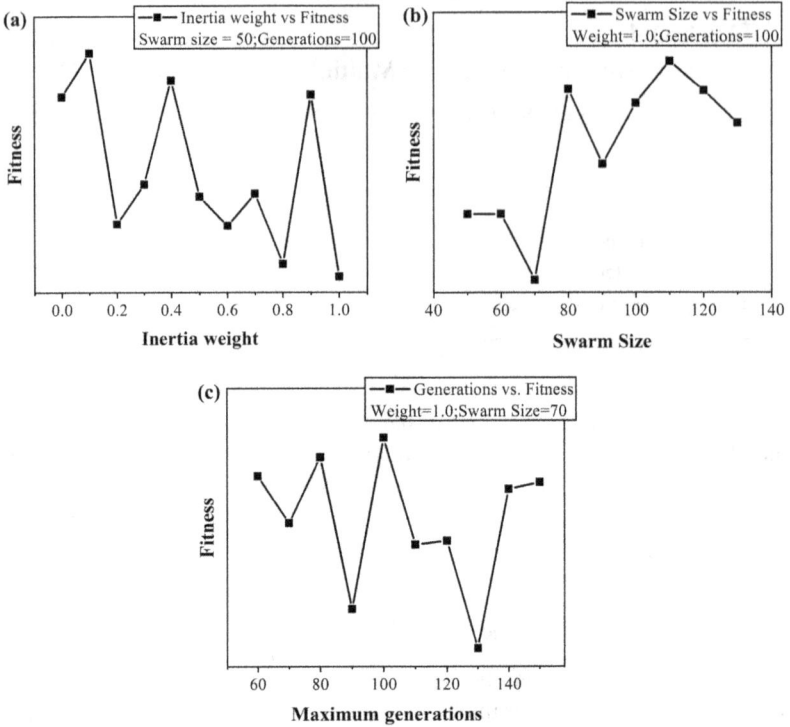

FIGURE 11.9　PSO parametric study: (a) inertia weight vs. fitness, (b) swarm size vs. fitness, (c) maximum number of generations vs. fitness.

TABLE 11.5
Optimum Hole Parameters for Multiple Responses Using PSO

		Optimum Values		
Process Parameters and Responses	Unit	Case 1 $(w_1 = 0.7; w_2 = 0.3)$	Case 2 $(w_1 = 0.6; w_2 = 0.4)$	Case 3 $(w_1 = 0.5; w_2 = 0.5)$
P: power	W	12	12	12
f: frequency	Hz	2	2	2
W: pulse width	%	6	6	6
D: standoff distance	mm	1.4967	1.738	1.3926
HD: hole diameter	μm	462.933	482.2867	500.7366
HAZ: HAZ thickness	μm	196.3462	200.3718	201.0977

In this case, three different cases were considered by giving various weight factors to the two process outputs, namely hole diameter and HAZ thickness. Table 11.5 shows the optimum performance of output parameters with different combinations of output parameters. As the aim is to minimize the objective function, case 1 is found to be suitable for a minimum hole diameter and HAZ thickness.

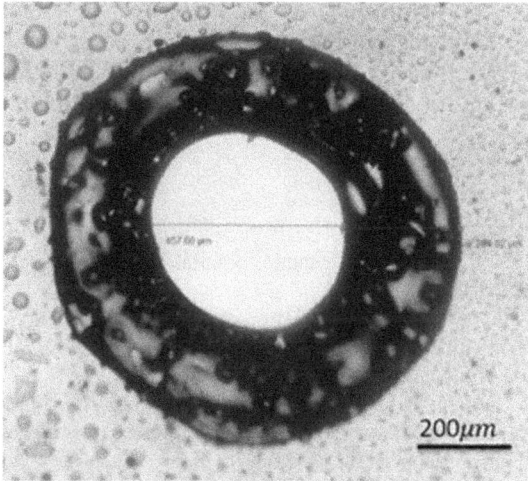

FIGURE 11.10   Micro-hole drilled with optimized parameters.

11.4.3   COMPARISON BETWEEN GA AND PSO

Several parameters need to be taken into account and specified during the optimization of a system using evolutionary algorithms as its speed of convergence is dependent on these parameters. The optimum parameter setting achieved while using GA
is as follows: 0.8 as the probability of crossover (P_c^*), 0.06 as the probability of
mutation (P_m), 80 as the population size (pop*), and 90 as the maximum number
of generations (G^*). In the case of PSO, the optimum parameters used are equal
to inertia weight (W^*) of 1.0, swarm size (SS^*) of 70, and the maximum number of
generations (G^*) of 130. The values of hole diameter and HAZ thickness obtained
experimentally are 457. 0 and 204.02 μm, respectively, using the optimized parameters of GA as depicted in Figure 11.10. The percentage error in values obtained for
HD and HAZ is 0.53% and 4.4%, respectively, for GA, while the percentage error in
the case of PSO is 1.2% and 3.7% for hole diameter and hole thickness, respectively.
It is interesting to note that the convergence rate is better in GA as compared to PSO.
Also, the error percentage for hole diameter is lesser when using GA. Moreover, the
convergence in GA was achieved irrespective of the weight percent applied to the
process variables in the objective function. The aim of the work being micro-drilling,
achieving a minimum hole diameter is of utmost importance.

11.5   CONCLUSION

In the present chapter, the contribution of nature-inspired algorithms toward the
optimization of advanced manufacturing systems was explored. The case of micro-
drilling on flexible substrates using a laser as the energy source was considered for
performing a multi-objective optimization. Two optimization techniques, specifi-
cally evolutionary algorithms, namely genetic algorithms (GA) and particle swarm
optimization (PSO), were explained in the course of this chapter. An attempt was

made to search for the optimized process parameter values for multi-objective functions, i.e., hole diameter and HAZ thickness of the drilled holes, using both GA and PSO. Genetic algorithm as the optimization technique was found to be suitable for the current process of micro-drilling with minimal error in the hole diameter obtained.

REFERENCES

1. Rosenberg RS. Simulation of genetic populations with biochemical properties: Technical report. 1967.
2. Lee TS, Ting TO, Lin YJ. An investigation of grinding process optimization via evolutionary algorithms. In *2007 IEEE Swarm Intelligence Symposium*, Honolulu, HI, USA, 2007 Apr 1 (pp. 176–81). IEEE.
3. Sardinas RQ, Santana MR, Brindis EA. Genetic algorithm-based multi-objective optimization of cutting parameters in turning processes. *Engineering Applications of Artificial Intelligence*. 2006; 19(2):127–33.
4. Datta R, Majumder A. Optimization of turning process parameters using multi-objective evolutionary algorithm. In *IEEE Congress on Evolutionary Computation, Barcelona, Spain*, 2010 Jul 18 (pp. 1–6). IEEE.
5. Hong X, Yuan L, Kaifu Z, Jianfeng Y, Zhenxing L, Jianbin S. Multi-objective optimization method for automatic drilling and riveting sequence planning. *Chinese Journal of Aeronautics*. 2010; 23(6):734–42.
6. Surekha B, Kaushik LK, Panduy AK, Vundavilli PR, Parappagoudar MB. Multi-objective optimization of green sand mould system using evolutionary algorithms. *The International Journal of Advanced Manufacturing Technology*. 2012; 58(1–4):9–17.
7. Hu X, Eberhart RC, Shi Y. Engineering optimization with particle swarm. In *Proceedings of the 2003 IEEE Swarm Intelligence Symposium, SIS 2003*, Indianapolis, IN, USA, (Cat. No. 03EX706), 2003 Apr 26 (pp. 53–57). IEEE.
8. Ali-Tavoli M, Nariman-Zadeh N, Khakhali A, Mehran M. Multi-objective optimization of abrasive flow machining processes using polynomial neural networks and genetic algorithms. *Machining Science and Technology*. 2006; 10(4):491–510.
9. Kuriakose S, Shunmugam MS. Multi-objective optimization of wire-electro discharge machining process by non-dominated sorting genetic algorithm. *Journal of Materials Processing Technology*. 2005; 170(1–2):133–41.
10. Rao RV, Pawar PJ, Shankar R. Multi-objective optimization of electrochemical machining process parameters using a particle swarm optimization algorithm. *Proceedings of the Institution of Mechanical Engineers, Part B: Journal of Engineering Manufacture*. 2008; 222(8):949–58.
11. Kant R, Gupta A, Bhattacharya S. Studies on CO_2 laser micromachining on PMMA to fabricate micro channel for microfluidic applications. In: Joshi S, Dixit U (eds.) *Lasers Based Manufacturing*. Springer, New Delhi, 2015. 221–38.
12. Kumar A, Gupta A, Kant R, Akhtar SN, Tiwari N, Ramkumar J, Bhattacharya S. Optimization of laser machining process for the preparation of photomasks, and its application to microsystems fabrication. *Journal of Micro/Nanolithography, MEMS, and MOEMS*. 2013;12(4):041203.
13. Dhupal D, Doloi B, Bhattacharyya B. Parametric analysis and optimization of Nd: YAG laser micro-grooving of aluminum titanate (Al_2TiO_5) ceramics. *The International Journal of Advanced Manufacturing Technology*. 2008; 36(9–10):883–93.
14. Biswas R, Kuar AS, Sarkar S, Mitra S. A parametric study of pulsed Nd: YAG laser micro-drilling of gamma-titanium aluminide. *Optics & Laser Technology*. 2010; 42(1):23–31.

15. Biswas R, Kuar AS, Biswas SK, Mitra S. Effects of process parameters on hole circularity and taper in pulsed Nd: YAG laser microdrilling of Tin-Al$_2$O$_3$ composites. *Materials and Manufacturing Processes*. 2010; 25(6):503–14.
16. Kibria G, Doloi B, Bhattacharyya B. Predictive model and process parameters optimization of Nd: YAG laser micro-turning of ceramics. *The International Journal of Advanced Manufacturing Technology*. 2013; 65(1–4):213–29.
17. Holland JH. *Adaptation in Natural and Artificial Systems*. University of Michigan Press, Ann Arbor, MI, 1975.
18. Kennedy J, Eberhart R. Particle swarm optimization. In *Proceedings of ICNN'95-International Conference on Neural Networks*, Perth, WA, Australia, 1995 Nov 27 (Vol. 4, pp. 1942–8). IEEE.

12 Modeling and Pareto Optimization of Burnishing Process for Surface Roughness and Microhardness

Vijay Kurkute
Department Mechanical Engineering
Bharati Vidyapeeth (Deemed to be University)
College of Engineering, Pune

Sandip Chavan
School of Mechanical Engineering
MIT World Peace University, Pune

CONTENTS

12.1 INTRODUCTION

The selection of manufacturing process is based on various factors. The important factors are manufacturing cost, machining time, accuracy, etc. But recently, an additional factor, functional performance, has become of significant importance. The main functional properties of engineered surfaces are physical, biological, and technological properties [1]. The technological properties include mechanical (fatigue and hardness) and tribological (friction and wear) properties. The functional performance is related to the surface generated by the manufacturing process. The functional performance, failure, and surface properties are interconnected [2]. The results of a survey [3] indicate that failure is correlated with surface properties. Approximately 25% of components fail due to fatigue [4]. Compressive residual stresses increase fatigue strength and hence reduce the chances of fatigue failure [5].

Burnishing, a post-machining process, is used to improve the external topography of surfaces (surface finish), microstructure, mechanical properties (microhardness), and residual stresses of the internal subsurface layer. The process is initially used in automobile components, including pistons, connecting rod bores, brake system components, transmission parts, and torque converter hubs. Nowadays, it is also used in non-automotive applications such as pistons, piston rods, and cylinders for the hydraulic or pneumatic system. In the hydraulic or pneumatic system, the surface is critical in the longevity and sealing properties. If the surface is too rough, oil is allowed to pass under the seal through the "valleys," causing oil to leak by the seal. A very smooth surface prevents the oil to pass under the seal. These increase the wear of the seal. For the lubrication of the seal, it is essential that a small amount of oil should pass under the seal. If the surface is too rough, an excess amount of oil passes under the seal, hence causing leakage. Hence, it is necessary to control the surface roughness. Controlling the surface roughness affects microhardness as well [6], which will affect the compressive residual stresses. It is observed that both the phenomena are contradictory to each other. Improving one affects the other.

The Mises–Hencky yield criterion is used to model the relationship between microhardness and residual stresses [6]. This model is adopted as a basis for the measurement of residual stresses in steel or other materials [5]. Hence, the study focuses on surface roughness and microhardness.

12.2 MOTIVATION

In the last two decades, researchers have investigated the effects of burnishing parameters and the surface modification produced by them, to address the issue of functional performance. The constant need for growing demand and functional performance is the driving force behind it [7]. Surface roughness, microhardness, and compressive residual stresses of the components produced in the burnishing process are critical from the product performance point of view. The four parameters: speed, feed, force, and the number of tool passes, are controlled in the burnishing process.

In the past, researchers have focused on the study, analysis, and optimization of surface roughness, microhardness, or compressive residual stresses. This leads to the optimization of one of the properties, which ignores its effect on other properties. Also, for the graphical analysis, it is essential to keep the two parameters constant and vary the remaining two. Such type of analysis is not useful in the situation where it is essential to control all the properties surface roughness, microhardness, and compressive residual stresses simultaneously. In this case, it is necessary to vary all the parameters simultaneously. Hence, it is required to establish a framework for multi-objective optimization of surface roughness and microhardness.

The multi-objective optimization plays an important role in our day-to-day life. Many scientific, social, economic, and engineering problems have parameters that can be adjusted to produce a more desirable outcome.

12.3 EXPERIMENT METHODOLOGY AND MODEL DEVELOPMENT

This work examines the effect of roller burnishing process on aluminum (Al 63400) workpiece. The optical emission spectroscope was used to verify the composition of the workpiece. The raw dimensions of the round workpiece were 32 mm diameter and 600 mm length. The initial turning operation was performed with speed. Initial turning conditions defined for all workpieces were as follows: cutting speed = 40 m/min, feed = 0.2 mm/rev, and depth of cut = 0.1 mm. The surface tester is used to quantify the surface roughness (Ra) value. The observed values were in the range of 1.7–2.18 μm. A single-roller carbide-burnishing tool was used in the experiment. The roller penetration was controlled by the spring fitted in the shank of the tool. The spring stiffness value of 19.33 N/mm was measured in the laboratory. This value was used for the calculation of the force.

In this work, four independent parameters which can be controlled during experiments are selected. These are speed, feed, force, and the number of tool passes. The response or dependent variable is microhardness. The range of controllable parameters was observed in trial experiments. After these trial experiments, the range for speed (20–50 m/min), feed (0.5–0.8 mm/rev), force (20–50 N), and the number of tool passes (1–5) was selected. The details of CCD model, experimental matrix, and quantification of responses are available in [8]. An empirical model is established for the surface roughness and microhardness. The model is developed using the RSM with CCD. The models were checked for statistical significance using various statistical tools, and the experimental runs were performed to confirm the model. The response surface analysis of both the models was performed. The models were optimized using a single-variable optimization technique to investigate the behavior around the optimum point.

The MOPSO algorithm was implemented in the MATLAB environment. The algorithm is used based on the non-dominated sorting method. The solutions obtained are known as a Pareto front. By using the Pareto front, it is possible to get a machining parameter for a particular combination of the surface roughness and microhardness.

Even if the machining process is the same, variations in the machining parameters result in different surface spectra. The spectrum depends on the application of the machining parameters used by the end user. Hence, the best way is to select a different combination of machining parameters and tools for a material grade. This exercise is performed for processes which are commonly used in the shop.

12.3.1 EMPIRICAL MODEL DEVELOPMENT FOR SURFACE ROUGHNESS AND MICROHARDNESS

In this study, the Design–Expert software is used to formulate the empirical response surface model. The relationships of the responses, surface roughness, and microhardness with the four independent variables speed, feed, force, and the number of tool passes were explored by using the least squares regression. The response surface model was then used to select an appropriate model. A fit summary statistic was used to select the appropriate model. It includes sequential model sum of squares (type I), lack of fit, and model summary statistics.

The p-value for the quadratic terms is less than 0.0001 for the surface roughness model. For microhardness model, the value is less than 0.0183. These small values indicate fit to the quadratic model. The p-values for lack of fit of surface roughness and microhardness are 0.0663 and 0.7732, respectively. These values are greater than the significance level $\alpha = 0.05$; hence, the lack of fit is not significant. The R^2 value for surface roughness and microhardness indicates that the model is expected to explain about 92% and 89% of the variability in predicting new observations. The cubic model has the highest R^2 value, but is aliased. This means that there are not enough unique design points to independently estimate all the coefficients for this model. Hence, the quadratic model is selected based on sequential model sum of squares (type I), lack of fit, and model summary statistics.

The solution is obtained with Design–Expert 7.0 software and is presented in Eqs. (12.1) and (12.2). It is the empirical model in coded form.

$$\text{Roughness} = 0.62 + 0.026x_1 + 0.47x_2 - 0.079x_3 + 0.048x_4 + 0.011x_1x_2 - 0.11x_1x_3$$
$$+ 0.084x_1x_4 - 0.049x_2x_3 + 0.16x_2x_4 + 7.062 10^{-3}x_3x_4 + 0.033x_1{}^2$$
$$+ 0.32x_2^2 + 0.078x_3^2 + 0.111 \tag{12.1}$$

$$\text{Microhardness} = 107.4885 - 1.3541x_1 - 2.0724x_2 + 7.3758x_3 + 8.7508x_4$$
$$- 3.4837x_1x_2 - 1.819x_1x_3 + 3.57x_1x_4 + 0.354x_2x_3$$
$$- 0.055x_2x_4 - 3.33x_3x_4 - 1.8600x_1^2 + 0.7299x_2^2$$
$$- 2.5125x_3^2 + 2.3436x_4^2 \tag{12.2}$$

12.3.2   The Development of Pareto Front

$$\text{Minimize or maximize } f_m(x) \ m = 1, 2, \cdots, M \tag{12.3}$$

$$\text{subject to } g_j(x) \geq 0 \ j = 1, 2, \cdots, J \tag{12.4}$$

$$h_k(x) = 0 \ k = 1, 2, \cdots, K \tag{12.5}$$

$$x_i^{\text{Lower}} \leq x_i \leq x_i^{\text{Upper}} \ i = 1, 2, \cdots, n \tag{12.6}$$

where
  $f_m(x) - M$ different objective functions.
  $g_j(x)$ and $h_k(x)$ – constraint functions.
  x_i^{Lower} and x_i^{Upper} – variable bound called decision space $\mathcal{D}$.

A solution X is a vector of n decision variables

$$X = \left[x_1, x_2, \cdots, x_n\right]'$$

Any solution X that satisfies all constraints and variable bounds is called a feasible solution. Solution X that does not satisfy all constraints and variable bounds are called an infeasible solution. Because of constraints, decision space $\mathcal{D}$ need not be feasible. The set of all feasible solution is called a feasible region $\mathcal{S}$.

For each solution $X = \left[x_1, x_2, \cdots, x_n\right]'$ there exists a point $f(x) = z = \left[z_1, z_2, \cdots, z_n\right]'$, a space known as objective space. Mapping takes place between n-dimensional decision space and m-dimensional objective space, as shown in Figure 12.1.

12.3.3   Pareto Optimal Solution

The concept of optimality for multi-objective optimization was formalized in 1900s in the field of economics; the credit goes to the Italian economist Vilfredo Pareto, who at the University of Lausanne in 1896 published the concept.

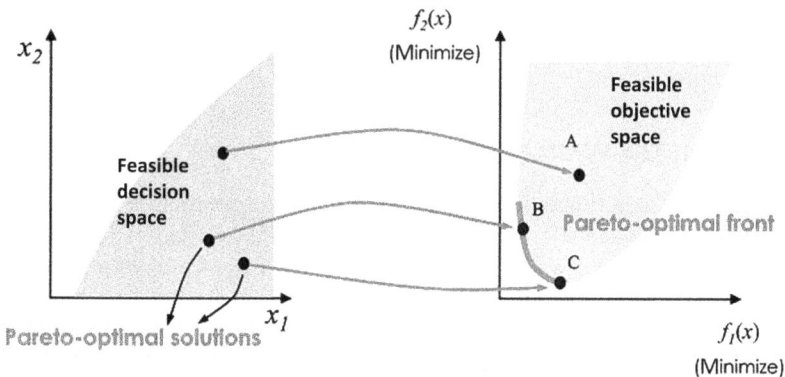

FIGURE 12.1   Design space and objective space.

Notations

$i \lhd j$ indicates that i is better than j on particular objective.
$i \rhd j$ indicates that i is worse than j on particular objective.
$i \ntriangleleft j$ indicates that i is not better than j on particular objective.
$i \ntriangleright j$ indicates that i is not worse than j on particular objective.

Definition 1

"A Pareto optimal solution is one for which any improvement in one objective will result in the worsening of at least one other objective; that is, a trade-off will take place. A dominated point is a point in the design objective space, for which there exists a point in the feasible space that is better (lower, in the case of minimization) in all objectives" [9].

A solution x^1 is said to dominate the other solution x^2, if the following conditions are true:

1. The solution x^1 is no worse than x^2 in all objectives. Thus, the solutions are compared based on their objective function values (or the location of the corresponding points z^1 and z^2).
   Mathematically, $f_m(x^1) \ntriangleright f_m(x^2)$ for all $m = 1, 2, \cdots, M$.
2. A solution $x^\wedge 1$ is strictly better than x^2 in at least one objective.
   Mathematically, $f_m(x^1) \lhd f_m(x^2)$ for at least one m belongs to $1, 2, \cdots, M$.
   Mathematically, $x^1 \prec x^2$ if x^1 dominates x^2.

For the second condition, there are two possibilities:

Strong dominance: $f_m(x^1) \lhd f_m(x^2)$ for all m belongs to $\{1, 2 ... M\}$

Weak dominance: $f_m(x^1) \lhd f_m(x^2)$ for atleast one m belongs to $\{1, 2 ... M\}$

Definition 2

(Non-dominated set). Among a set of solutions P, the non-dominated set of solutions P' are those that are not dominated by any member of the set P.

12.4 PARTICLE SWARM OPTIMIZATION

The particle swarm optimization (PSO) algorithm is a population-based search algorithm based on the simulation of the social behavior of birds within a flock. The initial intent of the particle swarm concept was to graphically simulate the elegant and unpredictable choreography of a bird flock. The method was introduced by Russell Eberhart, an electrical engineer, and James Kennedy, a social psychologist, in 1995 through the simulation of a simplified social model [10].

Imagine a situation of a group of birds searching for food in some region. In that region, there is only one place where food is located. All the birds do not know the position of the foodstuff. However, they know how far the food is in each iteration. So, what is the best approach to find the food? The efficient one is to adhere to the bird that is nearest to the food. The strategy of the group of the bird is used to solve the optimization problem. In PSO, every single solution is a "bird" in the pursuit zone. We call it the "particle." All of the particles have fitness costs, which are calculated by the fitness function to be optimized, and have velocities, which direct the flying of the particles. The particles travel through the problem space by observing the present optimum particles.

It starts with a randomly initialized population of the number of particles called the *swarm*. This population moves randomly through the search space. The particle remembers the best previous position of the self and its neighbor. Particles communicate the best position to each other. Based on the best position, particles adjust their position and velocity. In this way, particles fly toward better and better position in the search position. The process continues until the swarm move close to an optimum point.

12.4.1 Multi-objective Particle Swarm Optimization

The PSO strategy explained in the previous section is used to solve multi-objective optimization problems. The necessary changes should be incorporated in the original scheme. In case of multiple objective optimizations, solution set does not consist of a single solution. In multi-objective optimization, the solution is a set of different solutions called the Pareto optimal solution. In general, when dealing with a multi-objective problem, three main goals to accomplish are as follows [11]:

- Maximize the number of elements of the Pareto optimal set found.
- Minimize the distance of the Pareto front generated by the algorithm with reference to the true (global) Pareto front (presuming we know its location).
- Maximize the spread of solutions found, so that we can have a distribution of vectors as smooth and uniform as possible [11].

As we could see in the previous section, when solving single-objective optimization problems, the leader that each particle uses to update its position is completely determined once a neighborhood topology is established. However, in the case of multi-objective optimization problems, each particle might have a set of different leaders from which just one can be selected in order to update its position. Such set of leaders is usually stored in a different place from the swarm, which we will call external archive. This is a repository in which the non-dominated solutions found so far are stored. The solutions contained in the external archive are used as leaders when the positions of the particles of the swarm have to be updated. Furthermore, the contents of the external archive are also usually reported as the final output of the algorithm [12].

First, the swarm is initialized. Then, a set of leaders is also initialized with the non-dominated particles from the swarm. As we mentioned before, the set of leaders is usually stored in an external archive. Later on, some sort of quality measure is calculated for all the leaders in order to select (usually) one leader for each particle of the swarm. At each generation, for each particle, a leader is selected and the flight is performed.

Most of the existing MOPSOs apply some sort of mutation operator after performing the flight. Then, the particle is evaluated and its corresponding pbest is updated. A new particle replaces its pbest particle usually when this particle is dominated or if both are incomparable (i.e., they are both non-dominated with respect to each other).

After all the particles have been updated, the set of leaders is updated, too. Finally, the quality measure of the set of leaders is recalculated. This process is reiterated for a definite number of iterations.

12.4.2 Algorithm for MOPSO

The algorithm used in the current work is described here. The steps followed in the algorithm, with related MATLAB function, and burnishing responses, surface roughness and microhardness, are discussed.

12.4.2.1 Initialize the Population
The number of particles in the swarm influences the resulting performance. There is no definite rule for the selection of the number of particles (swarm size). The range used for various standard benchmarks functions is 20–100 [13,14]. All particles are assigned a random number. The boundaries in the current work range from [−2, −2, −2, −2] to [+2, +2, +2, +2]. Hence, all the particles are assigned with a random number in this range.

12.4.2.2 Initialize the Velocity
The velocity is initialized to zero, in all the directions of search space.

12.4.2.3 Evaluation of the Fitness
The objective functions are called fitness or cost function.

12.4.2.4 Best Fitness and Position
During the initialization stage, the random value of all particles and the corresponding fitness value are considered as best.

12.4.2.5 Non-dominated Points
The performance of the MOPSO is defined upon sorting the dominated and non-dominated points during the initialization and in each generation. The concept of external repository is introduced for possible storage of non-dominated points [15]. Initially, the repository is empty; hence, in the initialization phase, all non-dominated points are stored. The concept of external repository is discussed later.

12.4.2.6 Generate Hypercube
The object space is now divided into a number of hypercubes. Each particle located in the space is thus assigned to a particular unique hypercube. The edge length of the hypercube is calculated using the following formula:

$$\text{Edge length of hypercube} = \frac{f_{\max(x)} - f_{\min}(x)}{n\text{grid}} \tag{12.7}$$

$f_{\max(x)}$ – the maximum value of the fitness function.
$f_{\min}(x)$ – the minimum value of the fitness function.
$n$grid – the grid size.

Once the hypercubes are formed, the fitness value of each non-dominated point is fixed in the grid. Thus, the search space explored is represented as hypercubes in the objective space. From the set of non-dominated points, each point is assigned to a hypercube. Now, the hypercube contains $0,1,\cdots,n$ particles. From this, the quality of the hypercube is calculated using the formula [15]:

$$\text{Quality of hypercube} = \frac{10}{\text{number of particles in the hypercube}}$$

This aims to decrease the fitness of those hypercubes that contain more particles, and it can be seen as a form of fitness sharing.

12.4.2.7   Select Leader
The hypercubes are selected using roulette wheel selection with a probability that is directly proportional to the quality of the concerned hypercube.

12.4.2.8   Update Velocity
The velocity of the particles is updated in each generation.

12.4.2.9   Mutation Operator
PSO is known for its high convergence rate. This behavior may converge to local optimum, which leads to a false Pareto front. The use of mutation operator delays convergence and enables the particles to explore the whole search space. The mutation operator affects the entire population at the start of program execution. This intends to produce a highly explorative behavior in the algorithm. When the number of generations grows, the effect starts diminishing.

12.4.2.10   Maintain the Particles in Search Space
During the iterative process, particles may cross the search space. To avoid this, when the variable goes beyond the boundary, the following measures are taken:

1. If the particle is at the lower or upper boundary, then the particle's velocity is set to a value that corresponds to the value of the boundary.
2. At the boundary, the particle reverses its direction, as the same gets multiplied by −1.

12.4.2.11   Update Repository
The fitness of all the particles is evaluated in each iteration. The solutions, non-dominated points, are compared one at a time with the content of the repository. If the external archive is empty, then the current solution is accepted. If this new solution is dominated by an individual within the external archive, then such a solution is automatically discarded. Otherwise, if none of the elements contained in the

external population dominates the solution wishing to enter, then such a solution is stored in the external archive. If there are solutions in the archive that are dominated by the new element, then such solutions are removed from the archive. The repository capacity is initialized at the beginning of the generations; it is possible that the repository may get full as the number of generations increases. In such a situation, the adaptive grid mechanism takes control of the execution.

12.4.2.12   Update the Best Positions

The particle's position is compared with the positions in the repository. If it finds a better position, then its position is updated.

The flowchart of MOPSO is depicted in Figure 12.2.

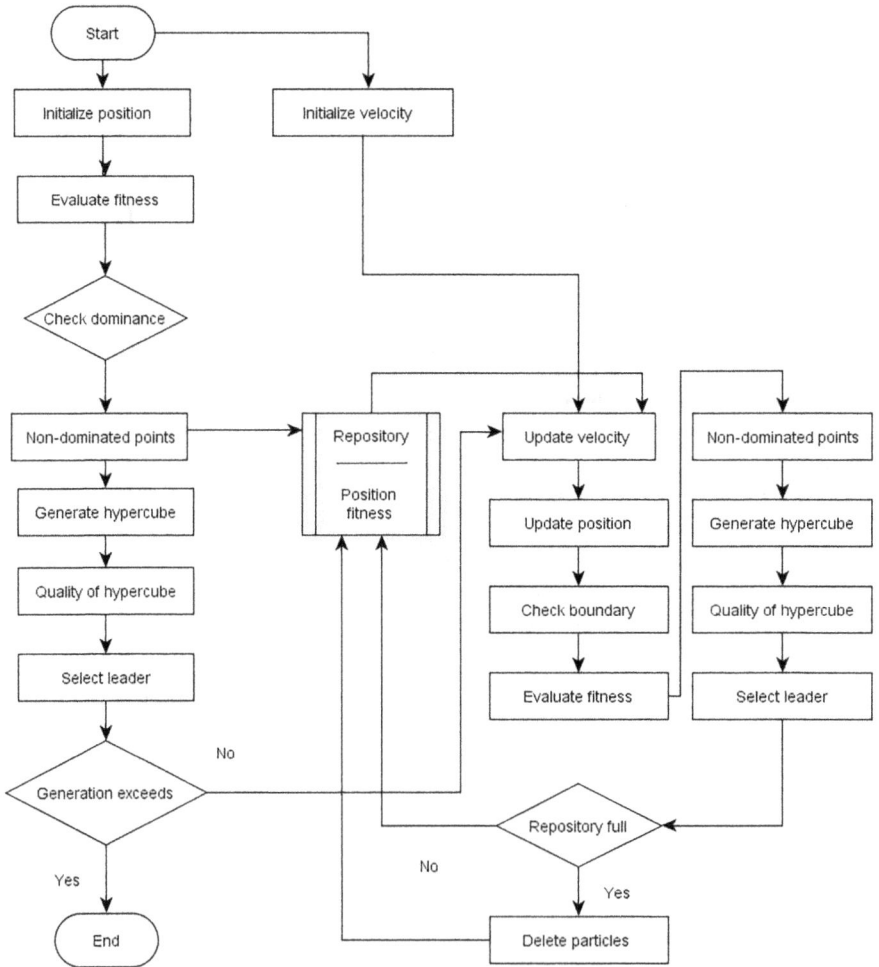

FIGURE 12.2   Flowchart of MOPSO.

12.4.3 MOPSO FOR SURFACE ROUGHNESS AND MICROHARDNESS

The following parameters are specified based on the previous studies in convergence and diversity of the Pareto front:

Population size =40; repository size = 100; maximum number of generations = 100; inertia weight = 0.3; individual confidence factor = 1.4; swarm confidence factor = 1.4; number of grids in each dimension = 5; maximum velocity in percentage = 5; and uniform mutation percentage = 0.3. With these parameters as input, the program is executed for population size 10, 20, 40, and 60. The resulting Pareto front is shown in Figures 12.3 and 12.4.

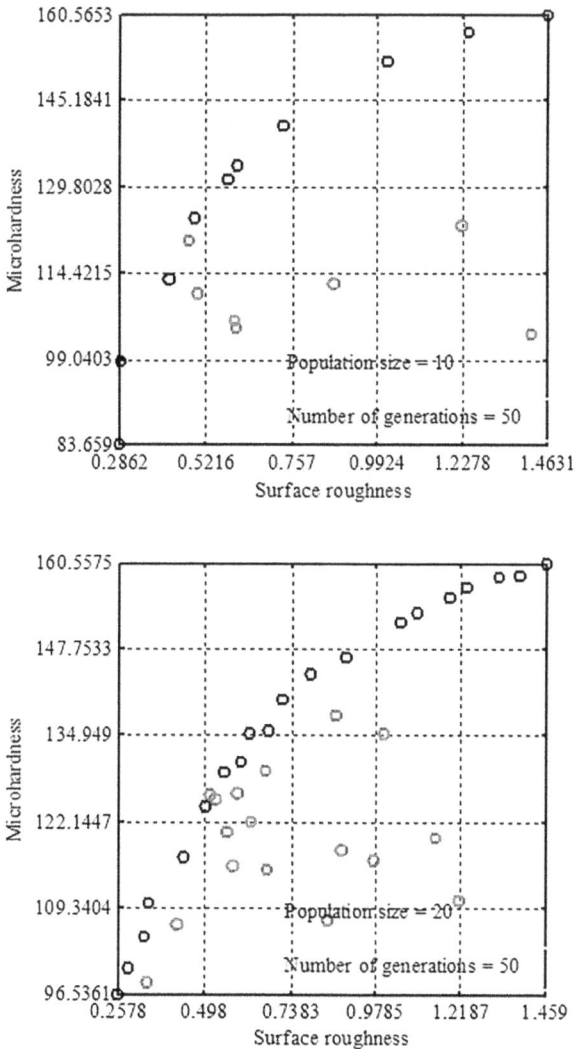

FIGURE 12.3   Pareto front for populations 10 and 20.

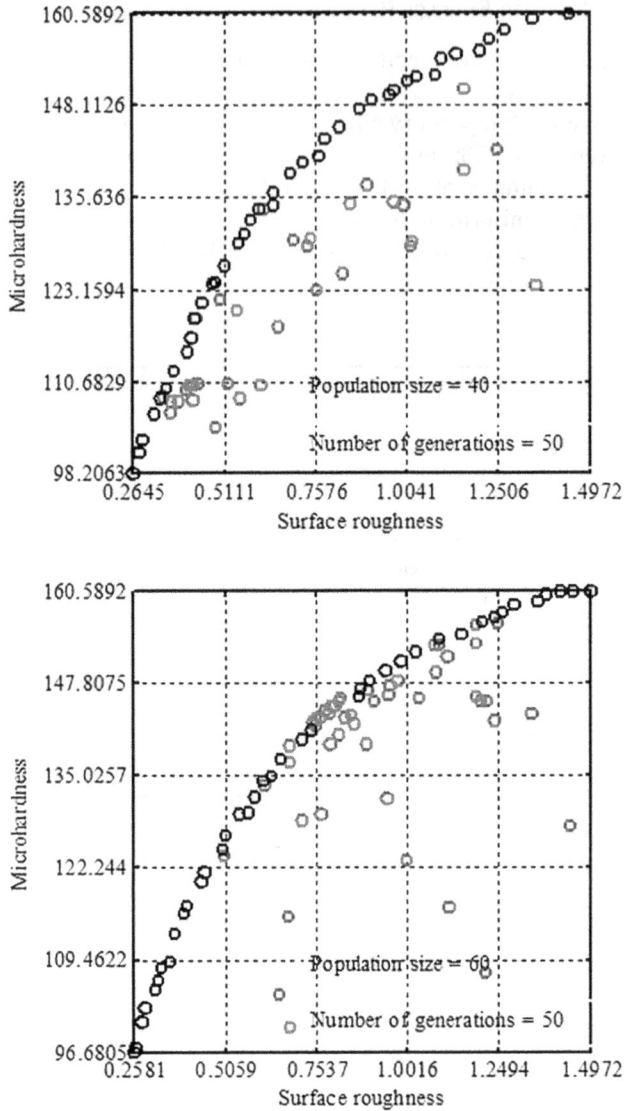

FIGURE 12.4   Pareto front for populations 40 and 60.

12.5   PERFORMANCE ASSESSMENT OF THE PARETO FRONT

Convergence criteria are discussed in [15,16]. The performance of a multi-objective optimization algorithm is verified using two goals. This is a quantitative assessment. The following are the goals that assess the performance [15]:

1. The Pareto front generated is compared with the known global Pareto front. The aim is to minimize the distance between these two fronts.

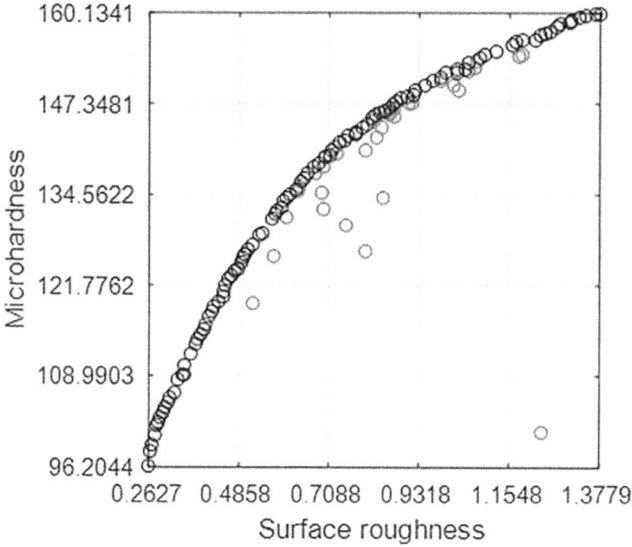

FIGURE 12.5 Performance of Pareto front.

2. The spread of the solutions in the Pareto front should be maximized. This ensures a uniform Pareto front.

The two goals are orthogonal to each other [17]. The first goal requires search toward the Pareto front, while the second goal requires search along the Pareto front, as shown in Figure 12.5.

The two goals are achieved using the following criteria:

• Metrics evaluating closeness to the Pareto front.
• Metrics evaluating diversity among non-dominated solutions.

12.5.1 Metrics Evaluating Closeness to the Pareto Front

In this method, the solutions obtained are compared with the known set of Pareto solutions. This is possible for the benchmark functions where the solution is already known. For benchmark functions, the error ratio is used. It was proposed by Van Veldhuizen [18]. The error ratio is defined as follows:

$$ER = \frac{\sum_{i=1}^{n} e_i}{n} \tag{12.8}$$

where n is the number of vectors in the current set of non-dominated vectors available. The ratio indicates the percentage of the non-dominated vectors found so far that are not the members of the true Pareto optimal set. $ER = 0$ indicates the ideal behavior.

In a practical problem, true Pareto front is always unknown; hence, Eq. (12.8) cannot be used directly. In such a situation, two ratios, improvement ratio and consolidation ratio, similar to the *ER* are investigated [19]. The archive at a generation is compared with an older archive. There are two situations:

1. The number of solutions in the old archive that are dominated by the newer archive (dominated solutions) is termed as *improvement ratio.*
2. The number of older archive members that are also present in the new archive (non-dominated solutions) is termed as *consolidation ratio.*

The improvement ratio and the consolidation ratio did not reach exact zero or unit values, respectively, due to numerical precision limitations. Nevertheless, a decreasing improvement ratio and/or an increasing consolidation ratio indicates the convergence [19]. The two ratios are presented in Figure 12.6. It shows that the archive population stabilized with evolution as the convergence was approached and the number of dominated solutions (improvement ratio) approached zero, whereas the sizes of new and old archives (consolidation ratio) became comparable; i.e., the consolidation ratio approached unity. It can be observed that after generation number 50, both the graphs stabilize.

12.5.2 METRICS EVALUATING DIVERSITY AMONG NON-DOMINATED SOLUTIONS

Spacing is one of the metrics used for evaluating diversity. This measures the spread of the non-dominated points. For this, the information about the "beginning" and "end" of the current Pareto front is used. Schott [20] suggested a metric which is calculated with a relative distance measure between consecutive solutions in the obtained non-dominated set, as follows:

$$S = \sqrt{\frac{1}{n-1}\sum_{i=1}^{n}\left(\bar{d}-d_i\right)^2} \tag{12.9}$$

where

$$d_i = \min_j \left(\left| f_1^i(x) - f_1^j(x) \right| + \left| f_2^i(x) - f_2^j(x) \right| \right),\ i,j = 1,2,\cdots,$$

$$\bar{n} = \frac{d_1 + d_2 + \cdots + d_n}{n}$$

n – the number of non-dominated points.

Equation (12.9) measures the standard deviation of different d_i values. If $S = 0$, then Schott suggests that the points on the Pareto front are equidistantly spaced [20]. The Deb [17] suggested that the standard deviation should be close to zero,

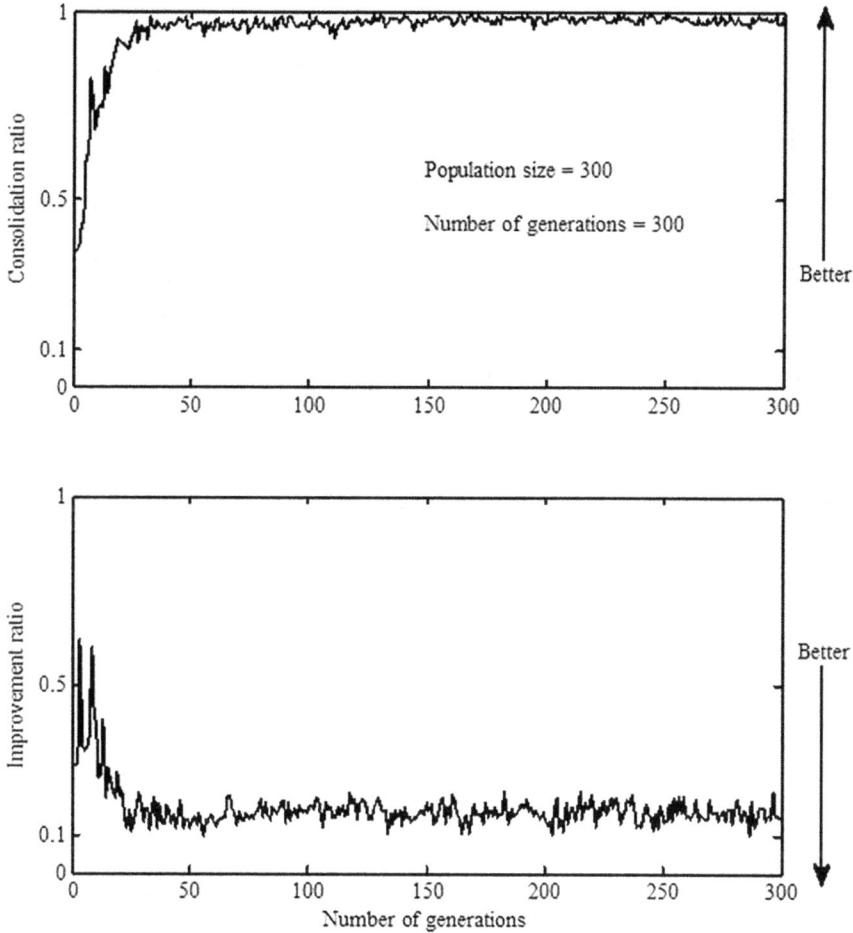

FIGURE 12.6   Metrics evaluating closeness to the Pareto front.

approximately between [0–0.5]. The spacing against each generation is depicted in Figure 12.7.

It shows the spacing up to 300 generations. The spacing ratio drastically falls from 1.0 to 0.1 in the first 50 generations. The value of S remains almost constant below 0.1 for generations 51–300. The ratio is just visible in the region 51–300 generations. The inset is presented in the same figure for 250–300 generations. It is clear that the ratio is oscillating at about 0.03.The values are approximately zero; hence, we can conclude that non-dominated solutions are uniformly distributed.

The percentage variation of the diagonal distance between extreme points against each generation is presented in Figure 12.7. The percentage variation in the diagonal length is very high in the generations 0–50. After that, the value remains constant at zero level as shown in the inset.

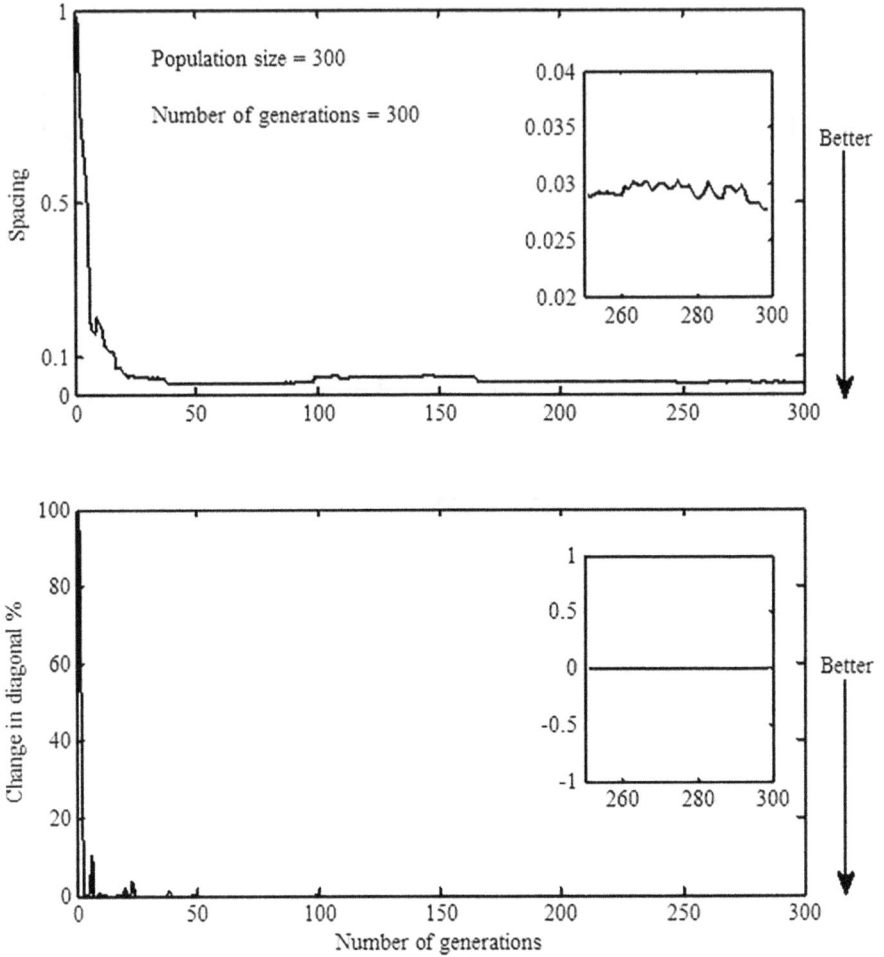

FIGURE 12.7 Metrics evaluating diversity among non-dominated solutions.

12.6 CONCLUSIONS

The important conclusions drawn from the investigations are as follows:

1. It was observed that an increase in the feed considerably affects the surface roughness. As feed is increased, the contact area between the tool and
   work surface increases due to which the asperities deform plastically. After
   achieving certain level in surface further increase in feed and number of
   passes distort the micro-profile and surface profile. It was observed that an
   increase in the force and the number of tool passes improves microhardness.
   With an increase in both the parameters, the tool penetrates beyond the
   maximum asperity height, which causes surface hardening.

2. To improve surface roughness, two factors the force and the number of tool passes can be held at a fixed value. For improvement in microhardness, the speed and feed are held at a constant value.
3. The above discussion confirms the contradictory behavior of the responses. The parameters speed and feed control the surface roughness, while the force and the number of tool passes control the microhardness. Hence, both the objectives cannot be improved simultaneously. A trade-off solution is obtained using the multi-objective optimization. It gives required machining parameters for a particular combination of surface roughness and microhardness.

REFERENCES

1. Bruzzone, A., et al., Advances in engineered surfaces for functional performance. *CIRP Annals*, 2008. 57(2): pp. 750–769.
2. Griffiths, B., *Manufacturing Surface Technology: Surface Integrity and Functional Performance*. 2001: Elsevier, London.
3. Tönshoff, H. and E. Brinksmeier, Determination of the mechanical and thermal influences on machined surfaces by microhardness and residual stress analysis. *CIRP Annals*, 1980. 29(2): pp. 519–530.
4. Findlay, S. and N. Harrison, Why aircraft fail. *Materials Today*, 2002. 5(11): pp. 18–25.
5. Buljak, V., et al., Assessment of residual stresses and mechanical characterization of materials by "hole drilling" and indentation tests combined and by inverse analysis. *Mechanics Research Communications*, 2015. 68: pp. 18–24.
6. Abbate, A., J. Frankel, and W. Scholz, Measurement and theory of the dependence of hardness on residual stress. 1993, Army apartment research development and engineering center watervliet ny benet labs.
7. Astakhov, V.P., Surface integrity–definition and importance in functional performance. In: Davim, J. (ed.) *Surface Integrity in Machining*, 2010: pp. 1–35. Springer, London.
8. Kurkute, V. and S.T. Chavan, Modeling and optimization of surface roughness and microhardness for roller burnishing process using response surface methodology for Aluminum 63400 alloy. *Procedia Manufacturing*, 2018. 20: pp. 542–547.
9. Messac, A. and A.A. Mullur, Multiobjective optimization: Concepts and methods. In: Arora, J.S. (ed.) *Optimization of Structural and Mechanical Systems*, 2007: pp. 121–147. World Scientific, Hackensack, NJ.
10. Eberhart, R. and J. Kennedy. A new optimizer using particle swarm theory. In: *Proceedings of the Sixth International Symposium on Micro Machine and Human Science, MHS' 95*, Nagoya, Japan. 1995. IEEE.
11. Coello, C.A.C, *An Introduction to Multi-Objective Particle Swarm Optimizers*. 2011: Springer, Berlin.
12. Devi, S., A.K. Jagadev, and S. Dehuri, Comparison of various approaches in multi-objective particle swarm optimization (MOPSO): Empirical study. In: Dehuri, S., Jagadev, A.K., and Panda, M. (eds.) *Multi-objective Swarm Intelligence: Theoretical Advances and Applications*, 2015:pp. 75–103. Springer, Berlin.
13. Bratton, D. and J. Kennedy. Defining a standard for particle swarm optimization. In: *Swarm Intelligence Symposium, SIS 2007*, Honolulu, HI, USA. 2007. IEEE.
14. Trelea, I.C., The particle swarm optimization algorithm: Convergence analysis and parameter selection. *Information Processing Letters*, 2003. 85(6): pp. 317–325.
15. Coello, C.A.C., G.T. Pulido, and M.S. Lechuga, Handling multiple objectives with particle swarm optimization. *IEEE Transactions on Evolutionary Computation*, 2004. 8(3): pp. 256–279.

16. van den Bergh, F., An analysis of particle swarm optimization. November, Ph. D. Dissertation, Faculty of Natural and Agricultural Sci., Univ. Petoria, Pretoria, South Africa, 2002.

17. Deb, K., *Multi-objective Optimization Using Evolutionary Algorithms*, Vol. 16. 2001: John Wiley & Sons, New York.

18. Van Veldhuizen, D.A. and G.B. Lamont, Multiobjective evolutionary algorithm research: A history and analysis. 1998, Citeseer.

19. Goel, T. and N. Stander, A study of the convergence characteristics of multiobjective evolutionary algorithms. In: *13th AIAA/ISSMO Multidisciplinary Analysis Optimization Conference*, Fort Worth, Texas, USA. 2010.

20. Schott, J.R., Fault tolerant design using single and multicriteria genetic algorithm optimization. 1995, Air Force Inst of Tech Wright- Patterson AFB OH.

13 Selection of Components and Their Optimum Manufacturing Tolerance for Selective Assembly Technique Using Intelligent Water Drops Algorithm to Minimize Manufacturing Cost

M. Siva Kumar, N. Lenin, and D. Rajamani
Vel Tech Rangarajan Dr. Sagunthala R&D
Institute of Science and Technology

CONTENTS

13.1 INTRODUCTION

The reliability, high quality, and low cost of products are the main requirements of customers in the competitive market. In this context, industries are looking for more suitable manufacturing processes to satisfy the customers. However, it is not possible to manufacture the products with exact dimensions due to the parametric constraints in all the manufacturing processes. Hence, an allowable variation is being considered from the nominal dimensions of the products. This allowable variation from the nominal dimension called tolerance plays a vital role in controlling the cost of the product. A wider tolerance reduces the manufacturing cost, whereas a tight tolerance increases the manufacturing cost. The distribution of tolerance among the components of the assembly called tolerance allocation also reduces the manufacturing cost. For making tight-tolerance components, it is necessary to go for secondary operations, which leads to an increase in the manufacturing cost. To avoid secondary operations, a technique called selective assembly is used, in which the subcomponents are manufactured with a wider tolerance, measured, and grouped into partitions and the corresponding group components are matched randomly. Manufacturing complex assemblies by combining tolerance allocation and selective assembly technique reduces the production cost to a large extent. Measurement of all components and surplus parts are the two main drawbacks which reduce its usage in the industry. Instead of making all components with a wider tolerance, particularly two components may be selected from the complex assembly to reduce the manufacturing cost. Further, the design of tolerance creates a link between the phases of design and manufacturing. The reliability of products is ensured by specifying the stringent tolerance. Usually, the manufacturing engineers like to have a wider tolerance for the economical manufacturing of the products. These requirements by the design and manufacturing engineers are being fulfilled by the selective assembly method.

13.2   RELATED RESEARCH

13.2.1   SELECTIVE ASSEMBLY

During the design process of the product and process, Kern [1] addressed the challenges and barriers in forecasting and managing the manufacturing variations. Furthermore, to overcome those barriers, he presented the tools and methods. Also, for various assembly techniques, the researcher developed closed-form equations to minimize the clearance variations. Mease et al. [2] considered few loss functions and dimensional distribution assumptions in optimal binning strategies to minimize the clearance variations of the products in an assembly. This strategy produced good results than the previous literature dealt with heuristic methods. Kannan et al. [3] proposed a new method for selective assembly. He used a genetic algorithm to achieve the minimum clearance variation by assembling components from different combinations of selective groups. This method took a radial assembly to analyze and identify the best combination. Kumar et al. [4] proposed a genetic algorithm to identify the optimum combination of component groups to minimize surplus parts while making assemblies. He considered a gearbox shaft assembly as an example case to show the efficiency of the proposed algorithm. Asha et al. [5] suggested a new selective assembly method to minimize the surplus parts and clearance variation by using the components piston and piston ring during the preparation of complex assembly. This proposed method used a non-dominated sorting genetic algorithm to find the best combination. Kannan et al. [6] proposed a particle swarm optimization algorithm to identify the selective group combinations for assembling the matching parts. This method reduced nearly 80% of assembly variations.

By considering the components with skewness in quality characteristics, Kannan et al. [7] suggested a new method for selective assembly. The main objective of this method was to minimize the clearance variations along with zero surplus parts. This method used a genetic algorithm to identify the number of components in the combinations of the selective group for a stated clearance variation. Wang et al. [8] proposed a new method for selective assembly to minimize the clearance variations of components in a gear assembly with zero surplus components using a genetic algorithm. Matsuura and Shinozaki [9] formulated an optimal manufacturing mean design to minimize the number of surplus components. The equal-width method was considered in this method to partition the components with a smaller variance in dimensions. Raj et al. [10] identified the optimal combination of all components to minimize the dimensional variations by using a genetic algorithm-based strategy. In the genetic algorithm, the length of chromosomes depended on the number of components. Hence, the limitation of this proposed method was the convergence, if the number of components of each assembly increases. Yue and Wu [11] demonstrated a genetic algorithm to make a hole and shaft assembly by identifying the optimal combination of selective groups with a minimum clearance variation. The clearance variation of the assembly by this method was low when compared to previous approaches. Babu and Asha [12] developed an artificial immune system algorithm to obtain the optimal combination of selective groups with the smallest amount of variation in the assembly tolerance and minimum loss value within the specification range. The evaluation of deviation from the mean was obtained using Taguchi's loss function method. Further, they analyzed the way to select the number of groups for selective assembly.

A vast literature is available on selective assembly considering reliable machines with infinite buffer capacity. But in the real case, in many assembly systems, unreliable machines and finite buffers are usually observed. Ju and Li [13] studied a two-component assembly system with unreliable Bernoulli machines and finite buffers. The system performance was analyzed using a two-level decomposition procedure. A high accuracy in performance evaluation was observed using the above-said method. Xu et al. [14] proposed a novel selective assembly strategy to improve the profit by minimizing the variation of components in the hard disk drive assembly. Discarding and binning theorems were formulated to discard the inferior components before assembly and to select the matching pairs of components. Lu and Fei [15] developed a selective assembly approach using a genetic algorithm to improve the success rate in manufacturing the assembly by reducing the surplus parts. A genetic algorithm with a specially designed 2D structure of the chromosome was used to achieve the objective. The proposed method was more suitable for the assembly with multiple dimension chains. Babu and Asha [16] proposed a symmetrical interval-based Taguchi loss function to evaluate the assembly loss in the selective assembly method. An improved sheep flock heredity algorithm was used to identify the good combination of the selective group to minimize the clearance variation and assembly loss value. Ju et al. [17] considered unreliable machines and finite buffers while manufacturing assemblies using a selective assembly method. Battery pack assemblies and power train production lines in the automotive industry were taken for this work. They assumed the Bernoulli machine reliability models. A two-level decomposition procedure was developed to evaluate the performance of the system. Liu and Liu [18] researched the selective assembly for the remanufacturing of engines. The number of groups and the range of each group were dynamic in the proposed work.

Chu et al. [19] developed a method for selective assembly to make the backlash of the RV reducer to meet the requirements. A mathematical model was established for this issue and solved using a genetic algorithm. Asha and Babu [20] introduced a metaheuristic method-based selective assembly technique to reduce clearance variation and surplus parts in complex assemblies such as ball bearings. Aderiani et al. [21] proposed a multistage approach to the selective assembly for minimizing clearance variation with no surplus parts. A hole and shaft assembly with two parts and a linear assembly with three parts were taken for this analysis. All dimensional distributions of the parts were considered. The best combination of selective groups was obtained by a genetic algorithm. An improvement of up to 20% in variation was achieved compared to the previous approaches.

13.2.2 INTELLIGENT WATER DROPS ALGORITHM

Numerous standard benchmark problems relevant to optimization have been solved using the intelligent water drops (IWD) algorithm. Hosseini [22,23] solved the traveling salesman problem using the IWD algorithm and proved the effectiveness of the IWD algorithm by comparing the results with the previous literature. The robot path planning solutions have been identified by Duan et al. [24,25] using the IWD algorithm. Again, Duan et al. [26] used the IWD algorithm for solving the queen puzzle and the multiple-knapsack problems and identified near-optimal solutions. Apart from that, various optimization problems related to various fields of study had been solved using the IWD algorithm and the results showed that the performance of the IWD algorithm was good compared to other well-known algorithms such as ACO and PSO. In power systems, the economic and emission dispatch problems were resolved by Abbasy and Hosseini [27], using the IWD algorithm. Similarly, Kamkar et al. [28] solved the vehicle routing problems, Alijla et al. [29] selected the features with rough sets, Hendrawan and Murase [30] selected the textural features for developing accurate irrigation system, and Hoang et al. [31] found the optimal data aggregation tree in wireless sensor networks using IWD algorithm. Niu et al. [32] solved the multi-objective job-shop scheduling problems using the IWD algorithm to identify the Pareto non-dominance schedules efficiently. Alijla et al. [33] introduced an ensemble of the IWD for feature selection related to human motion detection and motor fault detection.

13.2.3 INFERENCE FROM THE PAST WORKS

A considerable number of studies about selective assembly have been attempted by various researchers using different strategies with various optimization techniques. All the works focused on the minimization of clearance variation between components in an assembly and surplus components. However, the manufacturers are still looking for a better method to provide precisely assembled products at a low cost to gain more profit and hold their position in the competitive era. This work addresses a novel method for the above-stated purposes. Further, no one has attempted to solve selective assembly problems using an intelligent water drops algorithm. Hence, in this work, the intelligent water drops algorithm is proposed along with the novel strategy. The problem background and definition are detailed in the next section.

13.2.4 PROBLEM BACKGROUND AND DEFINITION

Globalization made the manufacturers to survive in the market with tough competition. With frequent changes in customer taste and technology, it is necessary for the manufactures to produce components with high quality and low price. Reducing the manufacturing cost is the only way to get a considerable profit in the globalized market. It is difficult to consider all the subcomponents of a complex assembly in the selective assembly since all subcomponents must be measured and grouped into the partition. And also, it is difficult and tedious to obtain the best combination of groups. A new method is introduced in this work to select logically two components from the complex assembly, which are selectively assembled based on the best group combination with other components manufactured as per allocated tolerance.

13.3 METHODOLOGY

The proposed method consists of four stages. In the first stage, the two best components and their tolerances (T_{iM}) are selected from the complex assembly based on % of manufacturing cost savings (MCS_i) using the existing (T_{iE}) and proposed tolerance of components (T_{iP}), as in Eq. (13.1). In the next stage, it is assumed that the selected components are manufactured with T_{iM} and the other components are manufactured with T_{iE} and randomly one thousand components are generated using "randn" MATLAB function and partitioned by equal-width (*EW*) method, in which the group width (gw_i) is calculated based on the number of groups (ng), as in Eq. (13.2). In the third stage, the assemblies are made by considering the assembly specification based on the best combination of groups obtained using the IWD algorithm. In the last stage, the surplus parts (*SP*) obtained in the previous stage are grouped by equal-area (*EA*) method based on the given group number (ng_{ea}). Usually, this group number falls between three and four. The assemblies are made by considering the assembly specification again based on the best combination of the group obtained using IWD for the equal-area method. The dimensional difference between the first (D_{ik}) and last component $\left(D_{NC_{ea},k}\right)$ falls in that group (*k*) called group width of equal-area (gw_{kea}) is calculated using Eq. (13.3) in which the value of gw_{kea} is depended on the number of components falls in the group (NC_{ea}).

$$MCS_i = 100 \times \frac{\left(MC_{iE} - MC_{iP}\right)}{MC_{iE}} \qquad (13.1)$$

where

$$MC_{iE} = C0_i e^{(-C1_i T_{iE})} + C2_i$$

$$MC_{iP} = C0_i e^{(-C1_i T_{iP})} + C2_i$$

$C0_i$, $C1_i$, and $C2_i$ – cost tolerance coefficients.

$$gw_i = \frac{T_{iM}}{ng} \qquad (13.2)$$

$$gw_{kea} = \left(D_{1k} - D_{NC_{eak}} \right) \qquad (13.3)$$

where

$$NC_{ea} = \frac{SP}{ng_{ea}}$$

The inspiration of water drops that flow in the river to reach the destination using the least path creates interest among the researchers to use the concept in the optimization problem. In this work, nodes (N_N) represent a sequence of numbers within the number of group "ng" of one component and a set of sequences of numbers corresponding to the number of components (N) represent one IWD (N_D).The total number of assemblies produced within the specification is considered as the objective, and the best combination of groups is obtained by implementing the IWD algorithm. The initialization of IWD parameters is listed in Table 13.1. Figure 13.1 illustrates the schematic diagram of the IWD algorithm.

13.4   NUMERICAL ILLUSTRATION

To illustrate the proposed methodology, a wheel mounting assembly shown in Figure 13.2 is considered in this work. Table 13.2 represents the existing allocated tolerance of the components and their exponential tolerance cost coefficients.

TABLE 13.1
Initialization of IWD Parameters

Particulars	Value
Velocity-updating parameters	$a_v = 1$; $b_v = 0.01$; $c_v = 1$
Soil-updating parameters	$a_s = 1$; $b_s = 0.01$; $c_s = 1$
Local soil-updating parameters	$\rho_n = 0.4$; $\rho_0 = 1-\rho_n = 0.6$
Global soil-updating parameters	$\rho_D = 0.8$; $\rho_s = 1+\rho_D = 1.8$
Initial soil (I_S)	10,000
Initial velocity (I_V)	200
Number of iterations (N_I)	100

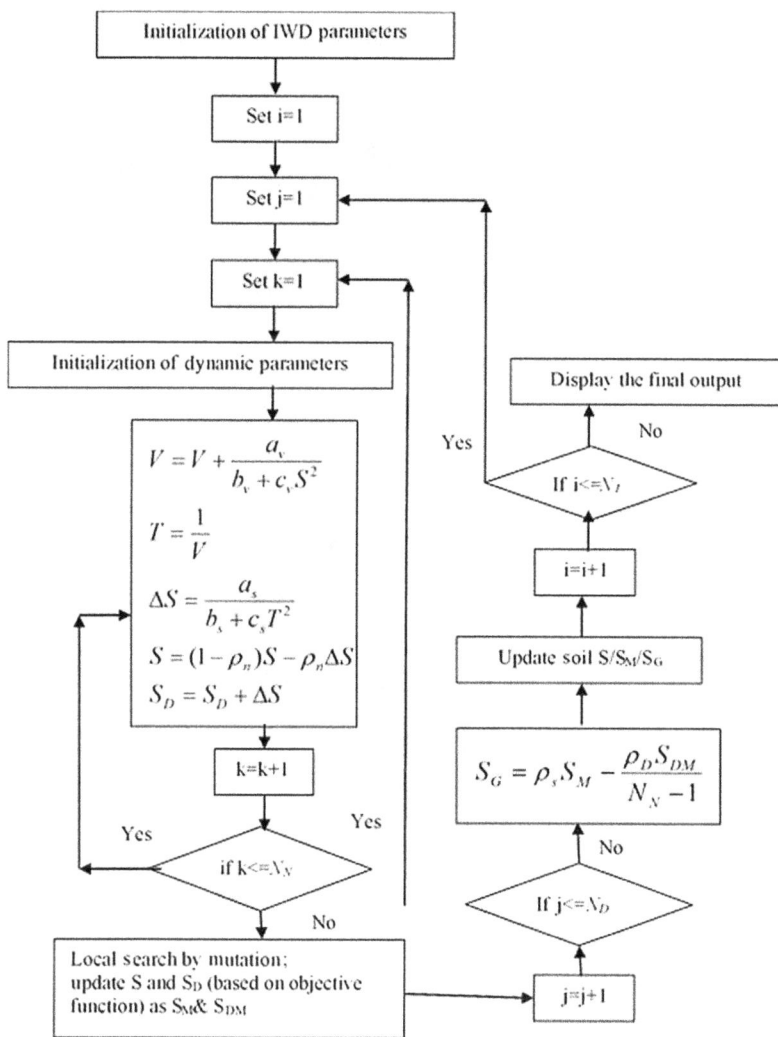

FIGURE 13.1 Schematic diagram of the IWD algorithm.

Figure 13.3 represents the tolerance cost relationship of components of WMA. As per stage 1 mentioned in the methodology, two components are selected using Eq. (13.1). Table 13.3 illustrates the manufacturing cost of the components by the existing and proposed methods and the manufacturing cost with *MCS*. It is understood that the components $X2$ and $X5$ have more *MCS* value than the other components; hence, these two components are selected for the selective assembly technique. It is assumed that components $X1$, $X3$, and $X4$ are bought-out components. As per the second stage, thousands of components $X2$ and $X5$ are generated using "randn" function, of which one component does not fall in the lower and upper specification limits; hence, in this work, it is proposed to produce 999 assemblies with the required specification,

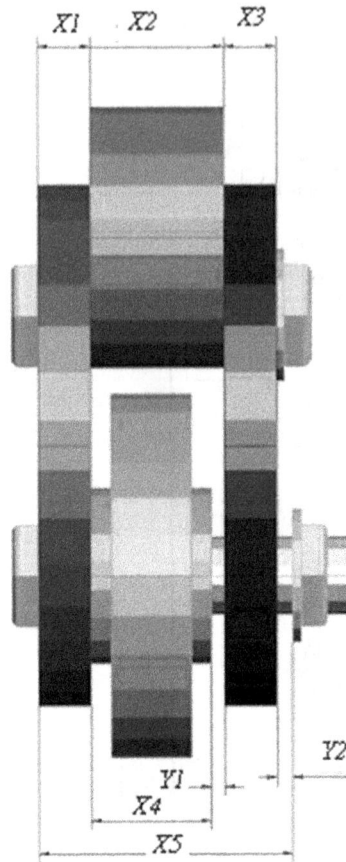

FIGURE 13.2   Wheel mounting assembly (WMA).

TABLE 13.2
Exponential Cost Function Constants for WMA

Part Name (i)	Dimension (mm)	$C0_i$	$C1_i$	$C2_i$	T_{iE} (mm)	T_{min} (mm)	T_{max} (mm)
$X1$	20	271.50	57.64	23.0	0.063	0.006	0.08
$X2$	50	271.50	57.64	23.0	0.056	0.006	0.08
$X3$	20	271.50	57.64	23.0	0.064	0.006	0.08
$X4$	45	352.43	92.70	35.0	0.054	0.002	0.06
$X5$	91	208.25	62.45	22.5	0.057	0.010	0.10

as given in Eq. (13.4). The dimensional distribution of $X2$ and $X5$ is depicted in Figure 13.4 based on different bin/group numbers (ng). Using Eq. (13.2), the group width of components $X2$ and $X5$ for various group numbers 5 and 6 is calculated and presented in Table 13.4.

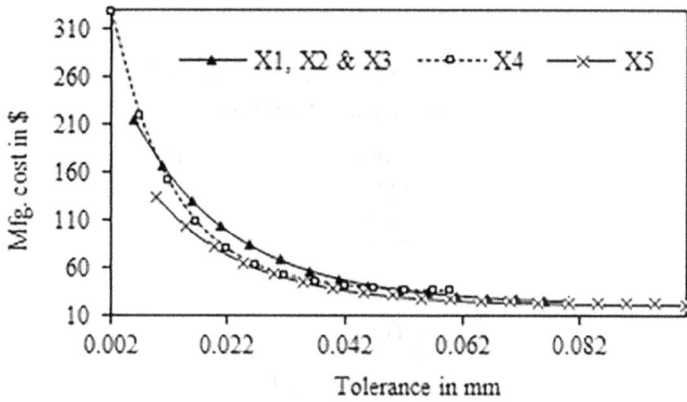

FIGURE 13.3 Tolerance cost curves of components of WMA.

TABLE 13.3
MCS and Manufacturing Tolerance (T_{iM}) of Components of WMA

Part Dimension	T_{iE}	MC_{iE}	T_{iP}	MC_{iP}	MCS_i	T_{iM}	MC_{iM}
X1	0.063	30.19	0.08	25.70	14.88	0.063	30.19
X2	0.056	33.76	0.08	25.70	23.89	0.080	25.70
X3	0.064	29.79	0.08	25.70	13.73	0.064	29.79
X4	0.054	37.36	0.06	36.35	2.70	0.054	37.36
X5	0.057	28.42	0.10	22.90	19.42	0.100	22.90
Total cost per assembly (PC_A)		159.52					145.94

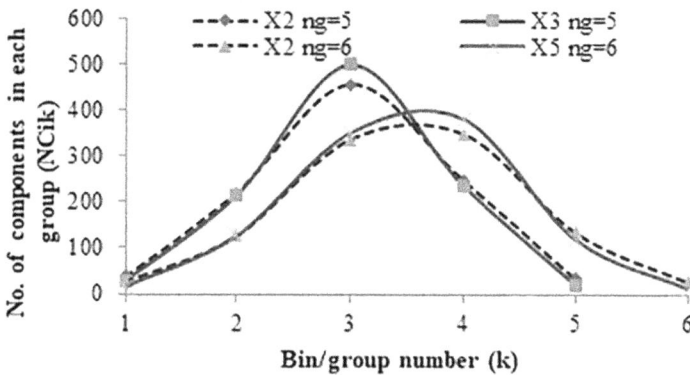

FIGURE 13.4 Dimensional distribution of X2 and X5.

TABLE 13.4
Group Width of Components X2 and X5 for Various Group Numbers

Ng	gw_{X2}	gw_{X5}
5	0.16	0.02
6	0.133333	0.016667

$$T_{Y1} = T_{X2} + T_{X4} \leq 0.11$$
$$T_{Y2} = T_{X1} + T_{X2} + T_{X3} + T_{X5} \leq 0.24 \tag{13.4}$$

The group width of the ith component for group number 1 to ng is calculated using Eq. (13.5), and the same for the components $X2$ and $X5$ is given for group numbers from 5 to 6 in Table 13.5.

$$gw_{ik} = gw_i * k \tag{13.5}$$

As per stage three, the best group combination is obtained by implementing the IWD algorithm. The step-by-step implementation procedure of the IWD algorithm for this problem is discussed below.

Step 1: The values of soil edge minimum (E_{min}) and maximum (E_{max}) are calculated using Eqs. (13.6) and (13.7), and the data are given in Table 13.1.

$$E_{min} = \frac{\rho_n * N_I}{b_s} \tag{13.6}$$

$$E_{max} = (\rho_s \rho_o)^{N_I} I_s \tag{13.7}$$

TABLE 13.5
Group Width of Components X2 and X5 for Various Group Numbers

	Group Width (gw_{ik})			
	X2		X5	
Group No. (k)	ng = 5	ng = 6	ng = 5	ng = 6
1	0.16	0.133333	0.02	0.016667
2	0.32	0.266667	0.04	0.033333
3	0.48	0.4	0.06	0.05
4	0.64	0.533333	0.08	0.066667
5	0.8	0.666667	0.1	0.083333
6		0.8		0.1

Step 2: In this case, nodes represent a sequence of numbers within the number of group "*ng*" of one component (shown in light gray color in Table 13.6) and a set of sequences of numbers corresponding to the number of components represent one IWD (shown in dark gray color in Table 13.6). For example, sample 1 is assumed as an initial solution. Two random numbers (*init_soil*) between E_{min} and E_{max} are generated along with a random number (R_c) between zero and one.

Step 3: To generate a good number of solutions, i.e., a combination of sequences of a group number, two methods, namely single-point and double-point crossover, are proposed based on the probability (P_c) value assumed as 0.4; i.e., if R_c is less than P_c value, then the single-point crossover method is adopted; otherwise, the double-point crossover method will be selected. Using Eq. (13.8), a cutting point (CP_{ij}) is computed. For example, the cutting point is 2 and the cutting method is the single-point crossover; then, the initial value of the first IWD is given below. In the initialization phase, the N_D number of sequences of the number of generated. For understanding purpose, *ng* is assumed as 5 and N_D is assumed as 7. Table 13.7 represents the method of generating the new combination of group numbers from

TABLE 13.6
Representation of Nodes and One IWD

	Sample 1				Sample 2	
	Component 1 (X2)	Component 2 (X5)			Component 1 (X2)	Component 2 (X5)
ng	Node 1	Node 2	ng		Node 1	Node 2
5	1 3 4 2 5	3 5 1 2 4	6		4 5 1 2 3 6	1 4 2 3 6 5

TABLE 13.7
Example Showing the Method of Generating New Sequences

Method 1 – Single–Point Crossover

	Component 1					Component 2				
Initial solution	1	3	4	2	5	3	5	1	2	4
IWD$_i$	4	2	5	1	3	1	2	4	3	5

Method 2 – Double–Point Crossover

	Component 1					Component 2				
Initial solution	1	3	4	2	5	3	5	1	2	4
IWD$_i$	1	5	4	2	3	3	4	1	2	5

TABLE 13.8

Initialization of the IWD Algorithm

IWD$_i$	Node 1						Node 2			
1	1	3	4	2	5	2	5	1	2	4
2	2	4	5	1	3	1	2	3	4	5
3	1	2	3	4	5	2	3	1	4	5
4	1	3	4	2	5	1	3	2	4	5
5	1	2	4	5	3	2	1	3	4	5
6	5	2	3	4	1	4	5	3	1	2
7	4	5	3	2	1	3	2	1	4	5

the initial solution using the *init_soil* for both the single- and double-point crossover methods. Table 13.8 represents the initialization of IWD values.

$$CP_{ij} = 1 + \frac{(init\_soil - E_{\min}) * (ng - 1)}{(E_{\max} - E_{\min})} \qquad (13.8)$$

Step 4: Using the group number of components shown in Table 13.7 and Eq. (13.5), the tolerance values of T_{X2} and T_{X5} are calculated. These values are substituted in Eqs. (13.3) and (13.4), where the tolerance of T_{X1}, T_{X3}, and T_{X4} are taken from Table 13.3 corresponding to column T_{iM}. If the above-calculated value is within the limit of 0.11 and 0.24 mm, the number of assemblies (NA_{ew}) based on matching the components of the *k*th group from the best group combination is calculated using Eq. (13.9). The surplus parts (SP) based on the matching of the best group combination are obtained using Eq. (13.10). Similarly, all other IWD's objective values are calculated. The best combination of group and its corresponding NA_{ew} and SP are stored in a file.

$$NA_{ew} = \sum_{k=1}^{ng} \min\left(NC_{ik}\right)_{i=1,2\ldots N} \qquad (13.9)$$

$$SP = 1000 - NA_{ew} \qquad (13.10)$$

Step 5: The velocity of IWD (V), time (T), amount of soil removed by IWD (ΔS), updated soil (US), and updated soil IWD (S_D) are calculated for each node using the equations given in Figure 13.1. If the value of the US is less than $E_{\min}$ or more than $E_{\max}$, then it is recalculated using Eq. (13.11).

$$\text{If } US < E_{\min}$$

$$US = 2 * E_{\min} - US$$

$$\text{If } US < E_{\max} \qquad (13.11)$$

$$US = 2 * E_{\max} - US$$

Step 6: In local search, it is assumed to generate ten different combinations of *US* (updated soil) using "randperm" function in MATLAB.

Step 7: The steps from 3 to 4 are carried out for each permuted combination obtained in the above steps.

Step 8: The value of the *US, IWD,* and combination of group numbers corresponding to the maximum number of assemblies is considered as local best.

Step 9: The values of the *US* obtained in the above step are considered as *init_soil,* and the steps from 2 to 8 are repeated up to (N_D-1) times.

Step 10: The IWD value of the *US* and S_D that has the highest *NA* is considered as the maximum soil (S_M) and maximum soil IWD (S_{DM}).

Step 11: To calculate the global soil (S_G) value, Eq. (13.12) is used.

$$S_G = \rho_s S_M - \frac{\rho_D S_{DM}}{N_N - 1} \tag{13.12}$$

Step 12: The value of soil corresponding to the maximum number of assemblies obtained by using *init_soil,* local search, and global search is considered as initial soil (*init_soil*) for the next iteration.

Step 13: The steps starting from 3 to 12 are repeated up to the stopping criterion, either reaching a specific iteration or reaching a specific value or no improvement in the last specific number of iterations.

In the last stage, the surplus parts (*SP*) obtained from the third stage are grouped based on the equal-area method. The number of components falls in each group, and its group width is calculated using Eq. (13.3). Similar to the third stage, the IWD algorithm (steps 1–13) is implemented to obtain the best combination of group and assemblies (NA_{ea}) that are made within the specified limitations (T_{y1} and T_{y2}). In this proposed method, the total number of assemblies produced is calculated by summing up the assemblies made from both the EW and EA methods and is expressed in Eq. (13.13).

$$NA_T = NA_{ew} + NA_{ea} \tag{13.13}$$

13.5   RESULTS AND DISCUSSION

Table 13.9 represents the best combination of groups with the number of assemblies obtained by implementing the IWD algorithm for both the EW and EA methods. It is understood from Table 13.9 that when the value of *ng* is equal to five, then there is a chance of a greater number of components in each group (*k*) and in the meantime the group width is also more. If assembly specifications (T_{y1} and T_{y2}) are not met while making assembly by matching the components, then the surplus parts will also be more in this case. This is the reason to get more *SP* as compared to group number 6. With increasing group numbers from 5 to 6, both group width and the number of components that fall in each group reduced. Because of this, a greater number of assemblies are possible within specified limits. This is reflected in Table 13.9.

TABLE 13.9

Best Combination of Groups and Number of Assemblies Made in EA and EW Methods

Ng	Equal-Width Method			Equal-Area Method			
	NA_{ew}	X2	X5	NA_{ea}	X2	X5	NA_T
5	918	1 5 2 4 3	5 1 4 2 3	52	123	123	970
6	955	3 4 5 6 1 2	4 3 2 1 6 5	42	231	312	997

The proposed cost per assembly (PC_A) for both the existing and selective assembly methods is given in Table 13.4. The proposed cost for making 1000 assemblies is obtained by multiplying PC_A by 1000, and it is given in Table 13.10. The actual cost per assembly (AC_A) is calculated using Eq. (13.14). The percentage of cost saved by the proposed method is computed using Eq. (13.15). The performance of the IWD algorithm is shown in Figure 13.5.

TABLE 13.10

Percentage of Cost Savings

Method	PC_A	PC_{1000}	NA_T	AC_A	%Savings
The existing method	159.52	159,520	1000	159.52	0
The proposed method for $ng = 5$	145.94	145,940	975	149.68	6.17
The proposed method for $ng = 6$	145.94	145,940	997	146.38	8.24

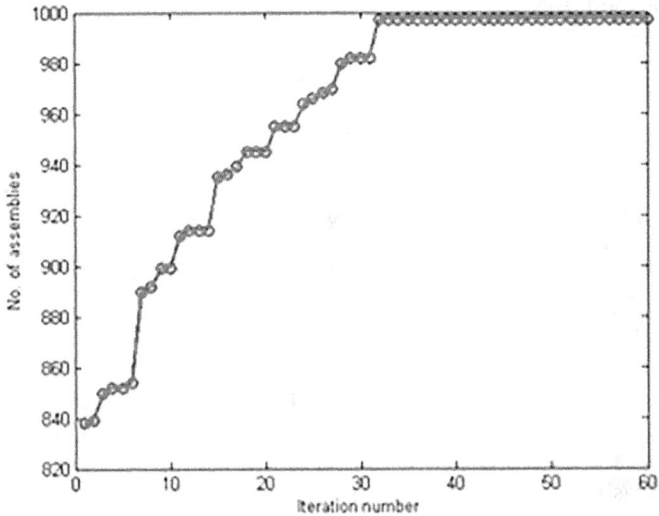

FIGURE 13.5 The performance of the IWD algorithm.

$$AC_A = \frac{PC_{1000}}{NA_T} \qquad (13.14)$$

$$\%\text{Saving} = 100 \times \frac{(PC_A - AC_A)}{PC_A} \qquad (13.15)$$

13.6   CONCLUSION

In this work, the manufacturing cost of complex assemblies was reduced using both the proposed equal-width and equal-area selective assembly techniques by implementing the IWD algorithm. To avoid tedious and laborious computation efforts, instead of considering all the components for selective assembly techniques, only two components were selected based on MCS. The number of groups plays a role in controlling surplus parts in the selective assembly technique. Two group numbers, i.e., group numbers 5 and 6, are considered to partition the components. The best results were obtained by manufacturing part dimensions $X2$ and $X5$ with T_{iP} and others with T_{iE} (allocated tolerances). The savings of 8.25% in the manufacturing cost per assembly is reported in this work by implementing the IWD algorithm to select the best group combination to match the components randomly from the selective groups. It is anticipated that the proposed work will be more beneficial to the manufacturing sectors. Further, the proposed work is more suitable for an assembly which contains any number of components with different dimensional distributions. The work can be extended further by considering the quality loss function while estimating the manufacturing cost of the assembly.

REFERENCES

1. Kern, D. C. (2003). Forecasting manufacturing variation using historical process capability data: Applications for random assembly, selective assembly, and serial processing. Ph.D. Thesis, Massachusetts Institute of Technology, USA.
2. Mease, D. N., Vijayan, N. & Sudjivnto, A. (2004). Selective assembly in manufacturing: Statistical issues and optimal binning strategies. *Technometrics* 46(2):165–175.
3. Kannan, S., Asha, A. & Jayabalan, V. (2005). A new method in selective assembly to minimize clearance variation for a radial assembly using genetic algorithm. *Journal of Quality Engineering* 17(4): 595–607.
4. Kumar, M. S., Kannan, S. & Jayabalan, V. (2007). A new algorithm for minimizing surplus parts in selective assembly by using genetic algorithm. *International Journal of Production Research* 45(20): 4793–4822.
5. Asha, A., Kannan, S. & Jayabalan, V. (2008). Optimization of clearance variation in selective assembly for components with multiple characteristics. *International Journal of Advanced Manufacturing Technology* 38(9–10): 1026–1044.
6. Kannan, S. M., Sivasubramanian, R. & Jayabalan, V. (2008). Particle swarm optimization for minimizing assembly variation in selective assembly. *International Journal of Advanced Manufacturing Technology* 42(7–8): 793–803.
7. Kannan, S. M., Sivasubramanian, R. & Jayabalan, V. (2009). A new method in selective assembly for components with skewed distributions. *International Journal of Productivity and Quality Management* 4: 569–589.

8. Wang, W., Li, D. & Chen, J. (2011). Minimizing assembly variation in selective assembly for complex assemblies using genetic algorithm. *Second International Conference of Mechanic Automation and Control Engineering (MACE)*, Hohhot, China, 1401–1406.

9. Matsuura, S. & Shinozaki, N. (2011). Optimal process design in selective assembly when components with smaller variance are manufactured at three shifted means. *International Journal of Production Research* 493: 869–882.

10. Raj, M. V., Sankar, S. S. & Ponnambalam, S. G. (2011). Genetic algorithm to optimize manufacturing system efficiency in batch selective assembly. *International Journal of Advanced Manufacturing Technology* 57(5–8): 795–810.

11. Yue, X. & Wu, Z. (2013). A heuristic algorithm to minimize clearance variation in selective assembly. *International Journal of Advanced Engineering Application* 6(3): 92–102.

12. Babu, J. R. & Asha, A. (2014). Tolerance modelling in selective assembly for minimizing linear assembly tolerance variation and assembly cost by using taguchi and AIS algorithm. *International Journal of Advanced Manufacturing Technology* 75: 869–881.

13. Ju, F. & Li, J. (2014). A bernoulli model of selective assembly systems. *Proceedings of the 19th World Congress,* The International Federation of Automatic Control Cape Town, August 24–29, 2014.

14. Xu, H. Y., Kuo, S. H., Tsai, J. W. H., Ying, J.F. & Lee, G.K.K. (2014). A selective assembly strategy to improve the components' utilization rate with an application to hard disk drives. *International Journal of Advanced Manufacturing Technology* 75(1–4): 247–255.

15. Lu, C. & Fei, J.F. (2014). An approach to minimizing surplus parts in selective assembly with genetic algorithm. *Institution of Mechanical Engineers, Journal of Engineering Manufacture* 229(3): 508–520.

16. Babu, J.R. & Asha, A. (2015).Modelling in selective assembly with symmetrical interval-based Taguchi loss function for minimising assembly loss and clearance variation. *International Journal of Manufacturing Technology and Management* 29(5/6): 288–308.

17. Ju, F., Li, J. & Deng, W. (2016). Selective assembly system with unreliable Bernoulli machines and finite buffers. *IEEE Transactions on Automation and Engineering* 14(1): 171–184.

18. Liu, S. & Liu, L. (2017). Determining the number of groups in selective assembly for remanufacturing engine. *Procedia Engineering* 174: 815–819.

19. Chu, X., Xu, H., Wu, X., Tao, J. & Shao, G. (2017). The method of selective assembly for the RV reducer based on genetic algorithm. *Journal of Mechanical Engineering Science* 232(6): 921–929.

20. Asha, A. & Babu, J.R. (2017). Comparison of clearance variation using selective assembly and metaheuristic Approach. *International Journal of Latest Trends in Engineering and Technology* 8(3): 148–155.

21. Aderiani, A.R., Wärmefjord, K. & Söderberg, R. (2018). A multistage approach to the selective assembly of components without dimensional distribution assumptions. *Journal of Manufacturing Science and Engineering* 140(7): 1015–1022.

22. Hosseini, S. H. (2007). Problem solving by intelligent water drops. *IEEE Congress on Evolutionary Computation*, Institute of Electrical and Electronics Computer Society, Piscataway, NJ, United States, Singapore, 3226–3231.

23. Hosseini, S. H., (2009). The intelligent water drops algorithm: A nature-inspired swarm-based optimization algorithm. *International Journal of Bio-Inspired Computation* 1: 71–79.

24. Hosseini, S. H., (2009). Optimization with the nature-inspired intelligent water drops algorithm. *Evolutionary Computation*, dos Santos, W.P. (ed.), I-Tech, Vienna, Austria, 297–320.

25. Duan, H., Liu, S. & Lei, X. (2008). Air robot path planning based on intelligent water drops optimization, *International Joint Conference on Neural Networks*. Institute of Electrical and Electronics Engineers Inc., Piscataway, NJ, USA, Hong Kong, China, 1397–1401.
26. Duan, H., Liu, S. & Wu, J. (2009). Novel intelligent water drops optimization approach to single UCAV smooth trajectory planning. *Aerospace Science and Technology* 13: 442–449.
27. Abbasy, A. & Hosseini, S. H. (2008). Ant colony optimization-based approach to optimal reactive power dispatch: A comparison of various ant systems, IEEE, Piscataway, NJ, USA, 282–289.
28. Kamkar, I., Akbarzadeh-T, M.R. & Yaghoobi, M. (2010). Intelligent water drops a new optimization algorithm for solving the vehicle routing problem. *IEEE International Conference on Systems, Man and Cybernetics*, IEEE, Piscataway, NJ, USA, 4142–4146.
29. Alijla, B.O., Peng, L.C., Khader, A.T. & Al-Betar, M.A. (2013). Intelligent water drops algorithm for rough set feature selection. *Fifth Asian Conference on Intelligent Information and Database Systems*, Part 2 ed. Springer Verlag, Kuala Lumpur, Malaysia, 356–365.
30. Hendrawan, Y. & Murase, H. (2011). Neural-intelligent water drops algorithm to select relevant textural features for developing precision irrigation system using machine vision. *Computers and Electronics in Agriculture* 77: 214–228.
31. Hoang, D.C., Kumar, R. & Panda, S.K. (2012). Optimal data aggregation tree in wireless sensor networks based on intelligent water drops algorithm. *IET Wireless Sensor Systems* 2: 282–292.
32. Niu, S.H., Ongn, S.K. & Nee, A.Y.C. (2013). An improved intelligent water drops algorithm for solving multi-objective job shop scheduling. *Engineering Applications of Artificial Intelligence* 26: 2431–2442.
33. Alijla, B.O., Lim, C.P., Wong, L.P., Khader, A.T. & Al-Betar, M.A. (2018). An ensemble of intelligent water drop algorithm for feature selection optimization problem. *Applied Soft Computing* 65: 531–541.

14 Enhancing the Surface Roughness Characteristics of Selective Inhibition Sintered HDPE Parts
An Integrated Approach of RSM and Krill Herd Algorithm

D. Rajamani, E. Balasubramanian,
and M. Siva kumar
Vel Tech Rangarajan Dr. Sagunthala R&D
Institute of Science and Technology

CONTENTS

14.1 INTRODUCTION

Additive manufacturing (AM) is a new class of layered manufacturing technique used for fabricating intricate customized components in short span and significantly reducing manufacturing cost [1]. Therefore, AM can be effectively used in fabricating net-shaped components in various fields such as aerospace, automotive, biomedical and defence industries [2]. In the past few decades, several AM techniques such as stereolithography, fused deposition modelling (FDM), selective laser sintering (SLS) and laser engineered net shaping (LENS) have been developed to fabricate different parts for various applications.

Each AM technique has its unique features such as processing capabilities, method of fabrication and diversity of raw materials. Among all these techniques, powder-based AM processes have recently been used in versatile applications of medium-to-high volume series production [3]. Direct metal laser sintering (DMLS), three-dimensional printing (3DP) and selective laser sintering (SLS) are the most widely used powder-based AM techniques in fabricating plastic, metal and ceramic parts. The utilization of high-cost heating mechanisms such as laser and electron makes the cost of the machine and processing high. Eliminating laser and electron heating elements used in AM processes has an incredible impression on reducing the machine cost and increasing the speed of the process.

In view of this, high-speed sintering (HSS) and selective inhibition sintering (SIS) processes are developed to significantly reduce the processing cost by eliminating the costlier heating elements. HSS involves a cost-effective infrared heater to sinter the powder particles. The heat is transferred to the powder surface through a radiation-absorbing material (RAM), which absorbs the as-received heat radiation from the infrared heater and transfers it to the powder particles for achieving effective sintering [4]. However, the incidence of RAM and high consumption of polymer powder materials are the foremost challenges to accustom HSS.

To overcome this issue, the SIS system was built at University of Southern California, USA [5]. In the SIS process, powder particles are sintered with desired part peripheries which are defined through precise delivery of inhibitors. The SIS has the key advantages of eliminating expensive tooling and support structure, processing various indigenously available raw materials such as polymers, metals and ceramics, which makes the process cost-effective.

Among the several advantages, SIS has few drawbacks such as compatibility issues with NC tool path generation, compaction of powder particles, less effective usage of raw materials and the lack of ability to improve the quality characteristics of sintered parts, such as strength, surface quality and dimensional stability [6]. The quality and performance features of the SIS parts can be improved by the appropriate selection of process parameters. Several SIS process parameters such as thickness of powder layer, supplied heat energy, part bed temperature, feed rate of heater and inhibitor printer and particle size have a prominent influence on the aforementioned functional qualities of the sintered parts [7]. Hence, it is necessary to investigate the influence of process parameters and identifying the optimal values of parameters to improve the quality of the sintered parts is very much essential.

Presently, various modelling and optimization techniques such as statistical (desirability, GRA and TOPSIS) and metaheuristic approaches (GA, SA, PSO, ABC, etc.) are used to enhance the performance of AM process and the quality of the fabricated parts [8–13]. However, solutions obtained using statistical optimization techniques are often discrete combinations of profound ranges of process parameters and they may fall in local optima [14]. Therefore, researchers are keenly looking for appropriate optimization technique to obtain global optimal solutions. However, several researchers have attempted the hybridization of optimization techniques to arrive at optimal process parameters [15–19].

As can be seen from the existing literature, there have not been any investigations on the implementation of novel metaheuristic optimization technique, namely krill herd algorithm (KHA), for the optimization of AM process. Moreover, due to the existence of several parameters and the non-linearity of the SIS process, KHA is adopted to identify the optimal parameters to enhance the surface quality of the sintered parts. The experiments are designed and executed by using RSM-BBD approach. The influence of the selected parameters on the surface roughness characteristics such as R_a, R_z and R_q is described with the aid of analysis of variance and interaction plots.

14.2 PROPOSED METHODOLOGY

14.2.1 Response Surface Methodology

Response surface methodology (RSM) [20–22] is employed for modelling and optimizing the governing parameters of the SIS. The Box–Behnken design (BBD) is a RSM-based approach, which is utilized to minimize the experimental runs. Quadratic models are established, and also the interactions between the governing parameters of the SIS are studied. The second-order polynomial equation formed from RSM is utilized to manifest the behaviour of the SIS process and is given in Eq. (14.1). In this investigation, the surface roughness characteristics of the sintered specimens are modelled accounting for the selected process parameters.

$$Y = C_0 + \sum_{i=1}^{n} C_i X_n + \sum_{i=1}^{n} d_i X_i^2 \pm \varepsilon \qquad (14.1)$$

Second-order quadratic models are developed using RSM for correlating the process variables and the responses. Additionally, ANOVA is exploited to justify the consequence of the developed quadratic models. The SIS process parameters and their levels are decided by conducting several pilot experiments and the existing literature. The selected SIS parameters and their lower and higher levels are described in Table 14.1.

14.2.2 Krill Herd Algorithm

Krill herd algorithm (KHA) is a population-based nature-inspired swarm intelligence metaheuristic technique enthused by the herding behaviour of krill individuals [23]. The objective function for KHA is formulated by the shortest distance between the

TABLE 14.1
SIS Parameters and Their Levels

S. No	Process Parameters	Unit	Range Low	Range Medium	Range High
1	Layer thickness	mm	0.1	0.15	0.2
2	Heat energy	J/mm^2	22.16	25.32	28.48
3	Heater feed rate	mm/s	3	3.25	3.5
4	Printer feed rate	mm/min	100	110	120

krill individuals and the location of food. In KHA, every individual krill is considered as the feasible solution for the optimization problem. The optimal solution is obtained in KHA by keeping the above goals in forward direction and the process of re-aggregation of individual krill. The general flow chart for the implementation of KHA is described in Figure 14.1.

The KHA proposed in this present investigation to obtain optimal SIS parameters to enhance the surface roughness characteristics comprises of the following steps:

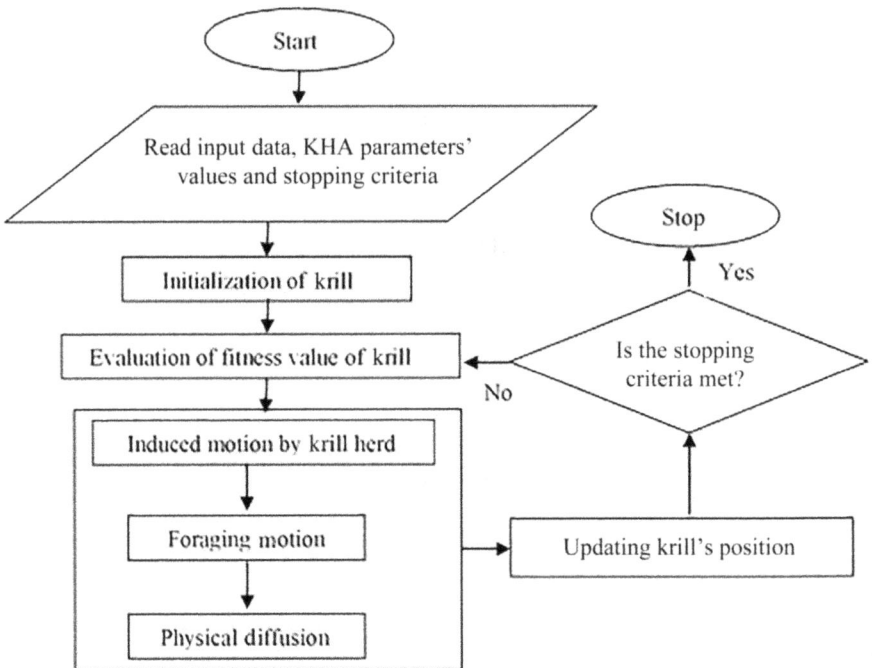

FIGURE 14.1 Flow chart of general KHA.

Step 1: Initialization: The algorithm parameters such as the number of krill individuals, stopping criteria and bounds of variables are initialized. The random values are assigned to D-dimensional individuals according to the following relation:

$$X_{i,j=0} = X_{i,\min} + \text{rand}_5 \times \left(X_{i,\max} - X_{i,\min} \right) \; i = 1,....,D; \; j = 1,....,NP \qquad (14.2)$$

where $X_{i,\min}$ and $X_{i,\max}$ represent the lower and upper levels of the *i*th decision variable, NP represents the population size, and rand$_5$ defines the uniformly distributed random variable between 0 and 1.

Step 2: Evaluation: According to the position and memories of the global best solution, every individual krill is evaluated.

Step 3: Motion calculation: In the search space, the time-dependent position of an individual krill is governed by the following three actions [24]:

i.   Induced motion by krill herd;
ii. Foraging motion; and
iii. Physical diffusion.

In general, an optimization algorithm should have the capability of searching spaces of arbitrary dimensionality. Hence, the following Lagrangian model is generalized to an *n*-dimensional decision space:

$$\frac{dX_i}{dt} = N_i + F_i + D_i \qquad (14.3)$$

where N_i is the motion induced by other krill individuals, F_i is the motion of foraging, and D_i is the physical diffusion of the *i*th krill individual.

The induced motion of the krill herd is subdivided into three components such as the target effect, the local effect and the repulsive effect. The motion of the krill can be represented by the following expression:

$$N_i^{\text{new}} = N^{\max}\alpha_i + \omega_n N_i^{\text{old}} \qquad (14.4)$$

where α_i is the direction of motion $\left(\alpha_i = \alpha_i^{\text{local}} + \alpha_i^{\text{target}} \right)$, ω_n is the inertia weight, $N^{\max}$ is the maximum induced speed, N_i^{old} is the last motion induced, and α_i^{local} and α_i^{target} are the local and target effects, respectively.

The foraging motion $\left(F_i \right)$ can be represented in two parts such as food location and its previous experience. The foraging motion of the *i*th krill can be expressed as follows:

$$F_i = V_f \beta_i + \omega_f F_i^{\text{old}} \qquad (14.5)$$

where $\beta_i = \beta_i^{\text{food}} + \beta_i^{\text{best}}$, V_f is the foraging speed, ω_f is the inertia weight, and F_i^{old} is the last foraging motion.

The physical diffusion (D_i) is assumed to be a random process, and it is expressed with the maximum diffusion speed (D^{max}) and a random directional vector (δ).

$$D_i = D^{max}\delta \qquad (14.6)$$

Step 4: Updation of krill's position: An individual krill's position in the search space is updated through implementing crossover and mutation functions.

Step 5: The algorithm will be stopped, and optimal solutions will be displayed if the stopping criterion is achieved; else, *Step 2* is repeated until the goal is attained.

14.3 EXPERIMENTAL DETAILS

A custom-built SIS system as shown in Figure 14.2 is developed to fabricate plastic near-net-shaped end-use components. The SIS system consists of the following elements: powder feed chamber which supplies the raw powder to the build chamber; part build chamber to fabricate the desired parts; a recycling chamber to collect the excess powder; an infrared heating element to sinter the powder surface; and an inhibition mechanism to define and separate the part boundary. According to the proposed Box–Behnken experimental design, 29 test samples of dimensions $135 \times 35 \times 8\,mm$ are fabricated to assess the surface roughness characteristics such as the arithmetic mean deviation (R_a), average peak-to-valley height (R_z) and root-mean-square value (R_q).

The surface roughness characteristics are measured at the top surface of the fabricated test specimens by using a non-contact-type 3D universal profilometer, and the measured roughness values are presented in Table 14.2.

FIGURE 14.2 Custom-built SIS system.

TABLE 14.2
Measured Surface Roughness Characteristics Values

Run	A: Layer Thickness (mm)	B: Heat Energy (J/mm²)	C: Heater Feed rate (mm/s)	D: Printer Feed rate (mm/min)	Surface Roughness		
					R_a (μm)	R_z (μm)	R_q (μm)
1	0.1	22.16	3.25	110	21.4	49.5	20.2
2	0.2	22.16	3.25	110	22.9	49.6	23.1
3	0.1	28.48	3.25	110	13.5	22.6	14
4	0.2	28.48	3.25	110	17.6	31.3	14.1
5	0.15	25.32	3	100	24.1	49.8	20.6
6	0.15	25.32	3.5	100	22.6	49.8	23.6
7	0.15	25.32	3	120	19.2	42.4	21.9
8	0.15	25.32	3.5	120	32.6	48.3	22.9
9	0.1	25.32	3.25	100	17.9	34.1	17.5
10	0.2	25.32	3.25	100	22.6	62	19.5
11	0.1	25.32	3.25	120	23.8	52.4	18.9
12	0.2	25.32	3.25	120	21.9	36.9	20.4
13	0.15	22.16	3	110	26	52.7	19.9
14	0.15	28.48	3	110	14.4	24.7	16.9
15	0.15	22.16	3.5	110	27.9	54.5	28.4
16	0.15	28.48	3.5	110	24.1	29.5	15.1
17	0.1	25.32	3	110	24.9	37.8	19.1
18	0.2	25.32	3	110	21.6	45.3	19.3
19	0.1	25.32	3.5	110	27.8	48.5	20.9
20	0.2	25.32	3.5	110	33.9	44.2	24.6
21	0.15	22.16	3.25	100	19.2	54.8	23.9
22	0.15	28.48	3.25	100	13.9	31.1	12.7
23	0.15	22.16	3.25	120	20.4	54.8	21.8
24	0.15	28.48	3.25	120	12.6	30.2	16.6
25	0.15	25.32	3.25	110	17.1	53.8	18.3
26	0.15	25.32	3.25	110	17.8	56.6	18.8
27	0.15	25.32	3.25	110	16.5	52.8	18.2
28	0.15	25.32	3.25	110	17.7	51.4	19.6
29	0.15	25.32	3.25	110	16.3	51.3	18.5

14.4 RESULTS AND DISCUSSION

14.4.1 STATISTICAL ANALYSIS OF THE DEVELOPED MODELS

Second-order polynomial models are developed to examine the surface characteristics such as R_a, R_z and R_q with respect to various SIS process parameters. The established model is evaluated using analysis of variance (ANOVA) with 95% confidence interval in Design Expert 7™ software, and the outcomes are given in Table 14.3. It is evident that for $p < 0.05$, they achieved 95% assurance level. Also, the factors

TABLE 14.3

ANOVA Results – Model Validation

Source	SS	DOF	Mean Square	F-Value	Prob > F	R^2	Adj. R^2
			R_a				
Model	783.73	12	65.31	67.25	< 0.0001	98.06	96.6
Total	799.27	28					
Residual	15.54	16	0.97				
Lack of fit	13.69	12	1.14	2.47	0.1984		
Pure error	1.85	4	0.46				
			R_z				
Model	3073.12	10	307.31	74.76	< 0.0001	97.65	96.34
Total	3147.11	28					
Residual	73.99	18	4.11				
Lack of fit	55.06	14	3.93	0.83	0.6473		
Pure error	18.93	4	4.73				
			R_q				
Model	324.78	11	29.53	88.30	< 0.0001	98.28	97.17
Cor total	330.46	28					
Residual	5.68	17	0.33				
Lack of fit	4.42	13	0.34	1.07	0.525		
Pure error	1.27	4	0.32				

of determination (R^2) for R_a, R_z and R_q are also found to be 98.06, 97.64 and 98.28, respectively. Hence, the developed models can fit the actual data.

The quadratic regression models are as follows:

$$R_a = 1982.28 - 617.53 \times A - 2.77 \times B - 959.03 \times C - 6.15 \times D$$

$$+ 188 \times AC - 3.3 \times AD + 2.47 \times BC + 1.49 \times CD + 1294 \times A^2$$

$$- 0.125 \times B^2 + 110.36 \times C^2 + 0.086 \times D^2 \tag{14.7}$$

$$R_z = -1505.78 + 3478.66 \times A + 38.98 \times B + 414.6 \times C + 3.12 \times D + 13.61 \times AB$$

$$- 236 \times AC - 21.7 \times AD - 2095.14 \times A^2 - 0.89 \times B^2 - 57.21 \times C^2 \tag{14.8}$$

$$R_q = 391.23 - 97.99 \times A + 8.78 \times B - 184.36 \times C - 3.16 \times D - 4.43 \times AB$$

$$+ 70 \times AC - 3.26 \times BC + 0.05 \times CD - 0.08 \times B^2 + 40.36 \times C^2 + 0.09 \times D^2 \tag{14.9}$$

Moreover, the normal probability plots shown in Figure 14.3a–c indicate that the blunders are dispersed normally, which implies that the proposed models are legitimately accurate and acceptable. Hence, the established second-order mathematical equations are realistically satisfactory in place of the SIS process.

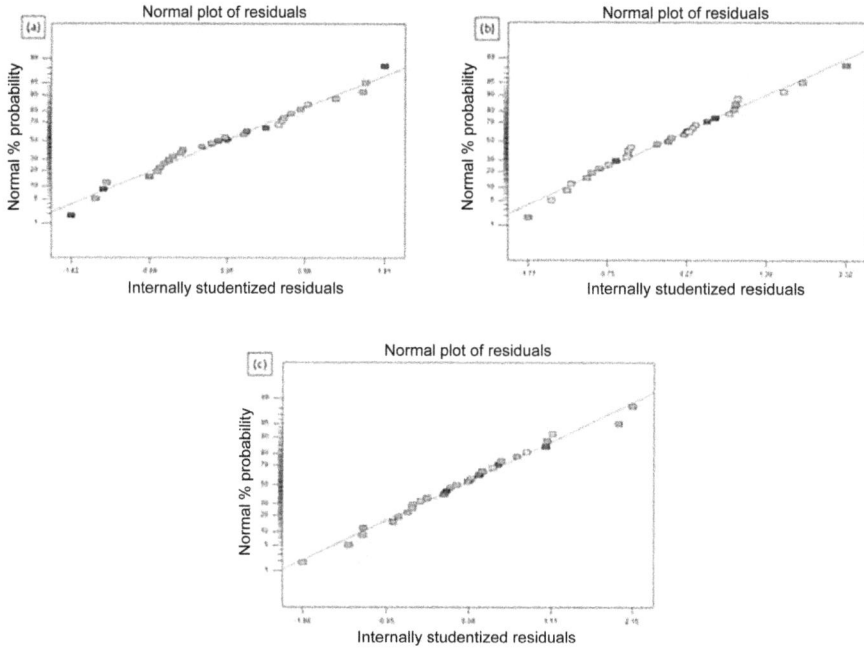

FIGURE 14.3 Normal probability plots for R_a, R_z and R_q.

14.4.2 Influence of Sintering Parameters on Roughness Characteristics

The enhancement of SIS-fabricated parts' surface quality is essential to improve the service life of the functional components. It is necessary to study the influence of various SIS parameters on examining the surface roughness characteristics. Thus, the interaction effects of various SIS process parameters obtained from RSM plots are shown in Figure 14.4a–h. These plots exhibit the relationship between two process variables, and the other two are kept at the centre point.

The effects of powder layer thickness and input heat energy on the surface roughness parameters are depicted in Figure 14.4 (a, d and g). From the interaction plots, it is found that both the selected parameters have a substantial influence on the surface roughness characteristics such as R_a, R_z and R_q. The surface roughness parameters are found to decrease with an increase in the applied heat energy, whereas an increase in layer thickness diminishes the surface quality. When the applied heat energy increases, the amount of heat supplied to the powder surface will increase. This sufficient heat energy facilitates strong inter-molecular bonding between the powder particles; hence, dense parts can be fabricated with improved surface quality [25]. An increase in the layer thickness causes poor heat penetration into the powder particles, resulting in thermal distortion in the sintered surfaces. Thus, the quality of part surfaces will diminish with increasing layer thickness.

Figure 14.4b, e and h portrays the variation in surface quality characteristics as a function of layer thickness and heater feed rate while the heater energy and printer

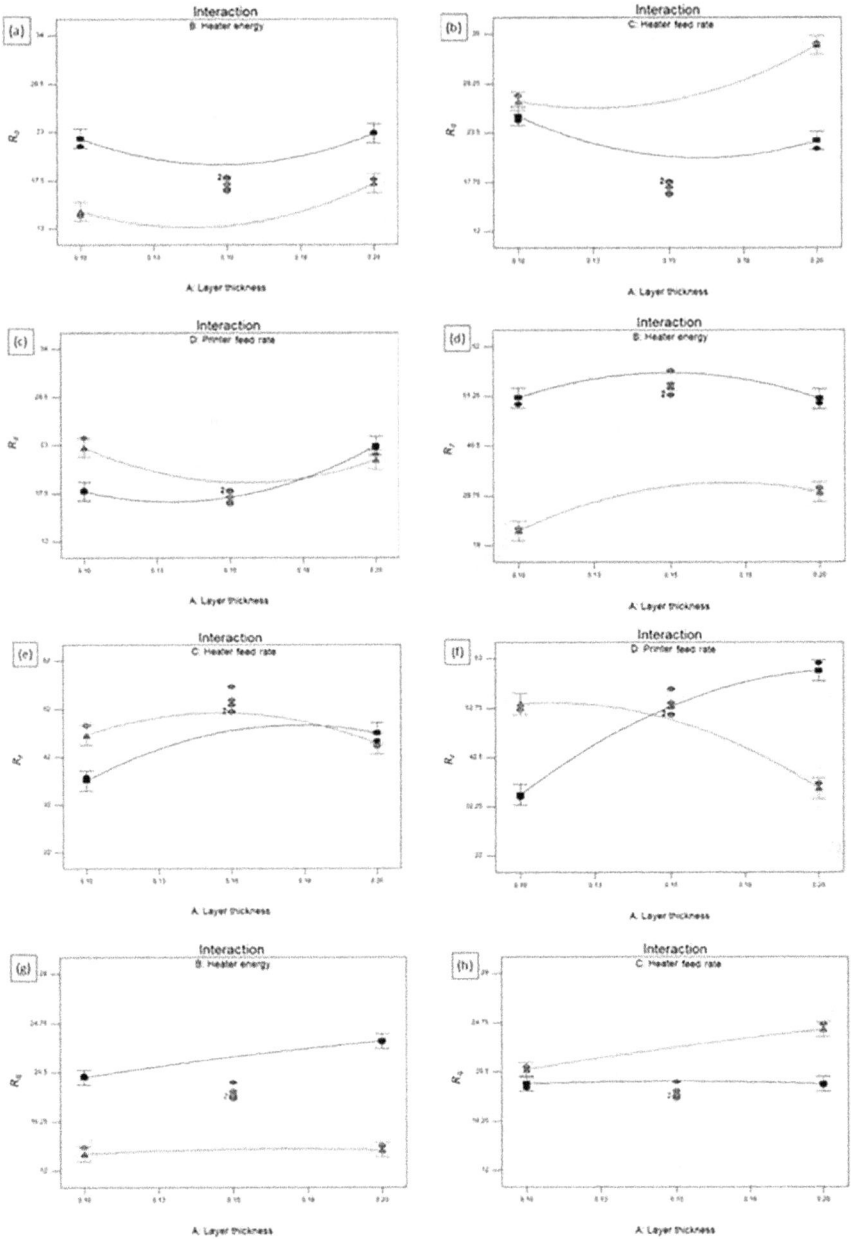

FIGURE 14.4 Interaction influence of SIS parameters on surface roughness characteristics.

feed rate are kept at middle range. It is inferred from the plots that the roughness parameters linearly increase from lower to higher levels with the increase in layer thickness and heater feed rate. As the feed rate of heater increases, the heat energy transferred to the polymer surface per unit time decreases due to the fast movement of heater set-up; hence, the polymer particles are sintered insufficiently and micro-voids and pores appear on the sintered surface. This leads to higher roughness of the sintered parts.

The combined effect of layer thickness and printer feed rate on the surface roughness parameters is depicted in Figure 14.4c and f. It can be seen that surface roughness increases up to a maximum level with an increase in the feed rate of printer and thickness of layer. This may be attributed to the fact that a maximum printer feed rate and layer thickness reduce the printing time on the boundary of part. Hence, it is difficult to separate the sintered and unsintered regions, resulting in higher surface roughness at the periphery of the sintered specimens.

14.5 MULTI-OBJECTIVE OPTIMIZATION USING KRILL HERD ALGORITHM

The objective of the present study is to minimize the surface roughness parameters such as R_a, R_z and R_q with respect to the selected bounds of SIS parameters. For this purpose, a swarm intelligence evolutionary optimization approach called krill herd algorithm (KHA) is used. The KHA code was developed and executed in MATLAB 2019b® environment. Second-order multiple linear regression models are utilized as objective functions for KHA to obtain optimal SIS parameters. Initially, the multi-objective functions are converted into single-objective functions using TOPSIS approach considering equal weights for each response. The closeness coefficient value of TOPSIS is considered as the objective function of KHA algorithm. The optimization problem of this present investigation is summarized as follows:

$$\text{Minimize} = \text{Roughness parameters}\left(R_a, R_z \text{ and } R_q\right)$$

subject to

$$0.1 \leq LT \leq 0.2$$

$$22.16 \leq HE \leq 28.48$$

$$3 \leq HFR \leq 3.5$$

$$100 \leq PFR \leq 120$$

The KHA is initialized with the parameters as given in Table 14.4 to obtain feasible optimal solutions to improve the surface roughness parameters of the SIS-processed parts.

The KH algorithm is allowed to run approximately 70 iterations for each trial to achieve meaningful optimal results and to avoid the objectives lying in local optima. Figure 14.5 shows the convergence plot of KHA for achieving higher closeness coefficient (objective of TOPSIS) values. From several iterations, the highest closeness coefficient of 0.9302 is achieved, which gives a layer thickness of 0.12 mm, a heater energy of 28.47 J/mm², a heater feed rate of 3.30 mm/s and a printer feed

TABLE14.4

KHA Parameters

Parameters	Notation	Value	Unit
Maximum induced speed	N_m	0.01	m/s
Inertia weight of the induced motion	ω_f	0–1	-
Positive constant to avoid singularities	Φ	0.2	-
Foraging speed	V_f	0.02	m/s
Inertia weight of foraging motion	ω_f	0–1	-
Maximum diffusion speed	D_m	0.005	-

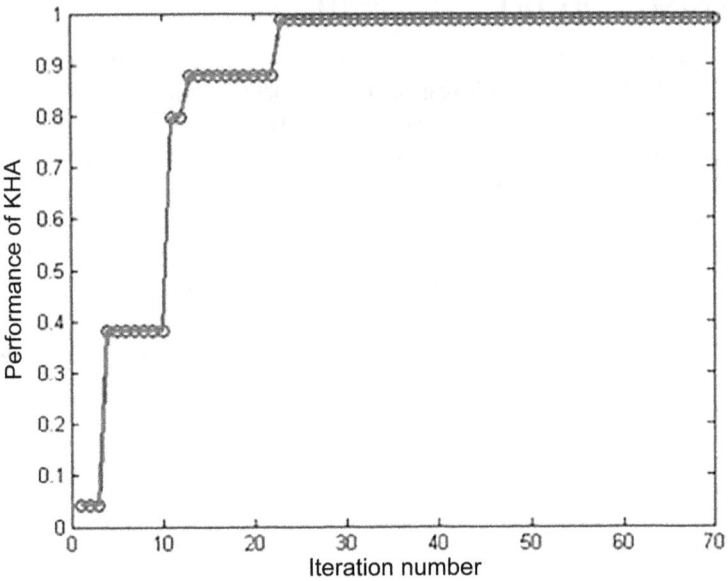

FIGURE 14.5 Convergence plot of krill herd algorithm.

rate of 100.90 mm/min as the optimal SIS process parameters to minimize the surface roughness characteristics as follows: R_a of 12.531 μm, R_z of 21.099 μm and R_q of 12.464 μm for the sintered high-density polyethylene parts.

14.6 CONCLUSION

The present work investigated the impact of the most influencing SIS process parameters on the surface roughness characteristics of the sintered high-density polyethylene parts. The optimal sets of process parameters were obtained using KHA metaheuristic technique. The following conclusions are drawn from the experimental and optimization studies:

- The SIS experiments were designed, and test specimens were fabricated through RSM-based BBD approach.
- The statistical analysis results showed that the heat energy and layer thickness are the most significant process parameters that influence the surface quality of the sintered parts followed by heater feed rate and printer feed rate.
- The surface quality of the sintered parts can be improved through increasing the applied heat energy and reducing the layer thickness, heater feed rate and printer feed rate.
- The RSM-KHA is found to be more efficient in predicting optimal SIS parameters through simplifying the optimization procedure.
- The optimal process parameters obtained through KHA are as follows: a layer thickness of 0.12 mm, a heater energy of 28.47 J/mm^2, a heater feed rate of 3.30 mm/s and a printer feed rate of 100.90 mm/min.

ACKNOWLEDGEMENT

The authors would like to sincerely thank the Armament Research Board, Defence Research and Development Organization, Government of India, for providing financial support to carry out this research work under the grant no. ARMREB/MAA/2015/167. The authors also thank The Director (R&D), Vel Tech Rangarajan Dr. Sagunthala R&D Institute of Science and Technology, for her continuous encouragement towards the research.

REFERENCES

1. Rajamani, D., & Balasubramanian E. (2019). Investigations of sintering parameters on viscoelastic behaviour of selective heat sintered HDPE parts. *Journal of Applied Science and Engineering* 22(3): 391–402.
2. Rajamani, D., & Balasubramanian E. (2019). Effects of heat energy on morphology and properties of selective inhibition sintered high density polyethylene. *Journal of Mechanical Engineering and Sciences* 13(1): 4403–4414.
3. Esakki, B., Rajamani, D., & Arunkumar, P. (2017). Modeling and prediction of optimal process parameters in wear behaviour of selective inhibition sintered high density polyethylene parts. *Progress in Additive Manufacturing* 3(3): 109–121.
4. Majewski, C. E., Oduye, D., Thomas, H. R., & Hopkinson, N. (2008). Effect of infrared power level on the sintering behaviour in the high-speed sintering process. *Rapid Prototyping Journal* 14(3): 155–160.
5. Khoshnevis, B., Asiabanpour, B., Mojdeh, M., & Palmer, K. (2003). SIS - a new SFF method based on powder sintering. *Rapid Prototyping Journal* 9(1): 30–36.
6. Rajamani, D., Balasubramanian, E., Arunkumar, P., Silambarasan, M., & Bhuvaneshwaran, G. (2018). Experimental investigations and parametric optimization of process parameters on shrinkage characteristics of selective inhibition sintered high density polyethylene parts. *Experimental Techniques* 42(6): 631–644.
7. Asiabanpour, B., Cano, R., Subbareddy, C., Wasik, F., VanWagner, L., & McCormick, T. (2007). A new heater design by radiation modeling and a new polymer waste-saving mechanism design for the SIS process. *Rapid Prototyping Journal* 13(3): 136–147.

8. Raju, M., Gupta, M. K., Bhanot, N., & Sharma, V. S. (2018). A hybrid PSO–BFO evolutionary algorithm for optimization of fused deposition modelling process parameters. *Journal of Intelligent Manufacturing* 30(7): 2743–2758.
9. Boschetto, A., Giordano, V., & Veniali, F. (2011). Modelling micro geometrical profiles in fused deposition process. *The International Journal of Advanced Manufacturing Technology* 61(9–12): 945–956.
10. Calignano, F., Manfredi, D., Ambrosio, E. P., Iuliano, L., & Fino, P. (2012). Influence of process parameters on surface roughness of aluminum parts produced by DMLS. *The International Journal of Advanced Manufacturing Technology* 67(9–12): 2743–2751.
11. Gholaminezhad, I., Assimi, H., Jamali, A., & Vajari, D. A. (2016). Uncertainty quantification and robust modeling of selective laser melting process using stochastic multi-objective approach. *The International Journal of Advanced Manufacturing Technology* 86(5–8): 1425–1441.
12. Mahapatra, S. S., & Sood, A. K. (2011). Bayesian regularization-based Levenberg–Marquardt neural model combined with BFOA for improving surface finish of FDM processed part. *The International Journal of Advanced Manufacturing Technology* 60(9–12): 1223–1235.
13. Strano, G., Hao, L., Everson, R. M., & Evans, K. E. (2011). Multi-objective optimization of selective laser sintering processes for surface quality and energy saving. *Proceedings of the Institution of Mechanical Engineers, Part B: Journal of Engineering Manufacture* 225(9): 1673–1682.
14. Javed, S., Mahmoudi, A., & Niazi, A. (2018). Investigation of drilling parameters on hybrid polymer composites using grey relational analysis, regression, fuzzy logic, and ANN models: A critical note. *Journal of the Brazilian Society of Mechanical Sciences and Engineering* 40(12): 560.
15. Peng, A., Xiao, X., & Yue, R. (2014). Process parameter optimization for fused deposition modeling using response surface methodology combined with fuzzy inference system. *The International Journal of Advanced Manufacturing Technology* 73(1–4): 87–100.
16. Panda, B. N., Bahubalendruni, M. V. A. R., & Biswal, B. B. (2014). A general regression neural network approach for the evaluation of compressive strength of FDM prototypes. *Neural Computing and Applications* 26(5): 1129–1136.
17. Vijayaraghavan, V., Garg, A., Lam, J. S. L., Panda, B., & Mahapatra, S. S. (2014). Process characterisation of 3D-printed FDM components using improved evolutionary computational approach. *The International Journal of Advanced Manufacturing Technology* 78(5–8): 781–793.
18. Mahapatra, S. S., & Sood, A. K. (2011). Bayesian regularization-based Levenberg–Marquardt neural model combined with BFOA for improving surface finish of FDM processed part. *The International Journal of Advanced Manufacturing Technology* 60(9–12): 1223–1235.
19. Sood, A. K., Ohdar, R. K., & Mahapatra, S. S. (2009). Parametric appraisal of fused deposition modelling process using the grey Taguchi method. *Proceedings of the Institution of Mechanical Engineers, Part B: Journal of Engineering Manufacture* 224(1): 135–145.
20. Tamilarasan, A., & Rajamani, D. (2017). Multi-response optimization of Nd: YAG laser cutting parameters of Ti-6Al-4V superalloy sheet. *Journal of Mechanical Science and Technology* 31(2): 813–821.
21. Ananthakumar, K., Rajamani, D., Balasubramanian, E., & Paulo Davim, J. (2019). Measurement and optimization of multi-response characteristics in plasma arc cutting of Monel 400™ using RSM and TOPSIS. *Measurement* 135: 725–737.

22. Rajamani, D., Ananthakumar, K., Balasubramanian, E., & Paulo Davim, J. (2018). Experimental investigation and optimization of PAC parameters on Monel 400™ superalloy. *Materials and Manufacturing Processes* 33(16): 1864–1873.
23. Gandomi, A. H., & Alavi, A. H. (2012). Krill herd: A new bio-inspired optimization algorithm. *Communications in Nonlinear Science and Numerical Simulation* 17(12): 4831–4845.
24. Wang, G.-G., Gandomi, A. H., Alavi, A. H., & Gong, D. (2017). A comprehensive review of krill herd algorithm: Variants, hybrids and applications. *Artificial Intelligence Review* 51(1): 119–148.
25. Rajamani, D., Ziout, A., Balasubramanian, E., Velu, R., Sachin, S., & Mohamed, H. (2018). Prediction and analysis of surface roughness in selective inhibition sintered high-density polyethylene parts: A parametric approach using response surface methodology–grey relational analysis. *Advances in Mechanical Engineering* 10(12): 1–16. doi:10.1177/1687814018820994.

15 Optimization of Abrasive Water Jet Machining Parameters of Al/Tic Using Response Surface Methodology and Modified Artificial Bee Colony Algorithm

K. Kiran
Dr. N.G.P Institute of Technology

K. Ravi Kumar
KPR Institute of Engineering and Technology

K. Chandrasekar
PSN College of Engineering and Technology

CONTENTS

15.1 INTRODUCTION

Metal matrix composites (MMCs) are gaining rapid importance in almost all engineering applications and industries. Abrasive water jet machining (AWJM) is a non-traditional process having tremendous advantages [1,2]. AWJM involves transformation of pressure energy of water into kinetic energy by passing it through a small opening (orifice) to perform the required operation. The abrasives are significant in machining, and the optimal recharging of abrasives enhances the penetration depth of water jets at a reduced cost [3,4]. Metallic materials machined by AWJ machining do not produce any heat-affected zone at the machining surface and have no thermal distortion. AWJM has the capability to machine hard and thick components with less cutting force and minimum stress compared to other machining processes. Machining performance is evaluated in terms of material removal rate, kerf dimensions and surface roughness influenced by machining parameters such as stand-off distance, traverse speed, jet pressure, nozzle diameter, jet angle and abrasive particle mix. Abrasive water jet machining is also used for shaping components under controlled conditions [5]. The combined effects of these process parameters largely affect the quality characteristics such as kerf taper angle and surface roughness. Abrasive water jet machining (AWJM) having high flexibility and reduced thermal expansion is one among the best machining techniques to machine complex shapes with the help of high jet energy containing abrasive particles [6–8]. The advantages of AWJM compared to other existing non-traditional process are as follows: (i) it covers more variety of materials to machine; (ii) there are no traces of heat-affected zones; (iii) machining forces on the top surface of the metal are negligible; and (iv) internal stresses and thermally altered layers in the cutting zone are absent [9,10]. AWJM can be used to machine a wide range of alloys and composites reinforced with hard ceramic particles having thickness ranging from 2 to 100 mm [11].

The machining process is measured in terms of more than one parameter, and hence, it becomes complex in carrying experiments involving various machining parameters. Several tools such as Taguchi's method, Taguchi grey relational analysis, artificial neural networks, weight method and desirability approach have been employed in developing mathematical models for predicting and optimizing the machining parameters in various engineering applications [12,13]. Response surface methodology is also one of the prominent tools used for designing the experiments, and the machining parameters can be analysed by ANOVA to check the validity of the developed model [14,15]. In addition to the empirical and mathematical models, models using soft computing techniques and hybrid approaches to solving multi-objective optimization problems have also been also developed [16–18].

Numerous optimization algorithms have broadly been used over the past decades, to solve manufacturing and structural optimization problems. Nature-inspired algorithms such as simulated annealing, genetic algorithms, differential evolution, ant and bee colony algorithms, particle swarm optimization, firefly algorithm, cuckoo algorithm, bat algorithm, harmony search and flower algorithm are the most commonly used algorithms for solving a variety of optimization problems [19]. Dervis Karaboga and Bahriye Basturk [20] developed an artificial bee colony (ABC) algorithm to solve multi-objective optimization problems. The ABC algorithm is based on the swarm behaviour and communication between three sets of bees searching for a good-quality food source. Researchers employed ABC algorithm to solve single- and multi-objective optimization problems and compared the results with other optimization methods [21]. The ABC algorithm was successfully used by researchers to optimize parameters in supplier integration, electrochemical machining, ultrasonic echo, manufacturing design problems, etc. [22–26].

A literature study reveals that abrasive water jet machining is one of the non-traditional machining processes that can be utilized to machine metals and composites successfully. AWJM is influenced by a number of machining parameters such as stand-off distance, traverse speed, jet pressure, nozzle diameter, abrasive particle mix and abrasive size. Mathematical models can be developed using response surface methodology to evaluate the relation between the input parameters and their outputs. Artificial bee colony algorithm, one of the nature-inspired swarm optimization techniques, can be utilized to solve multi-objective optimization problems. In this study, pure aluminium having 99% purity composites reinforced with different percentages of TiC is fabricated by the stir casting technique. An attempt is made to optimize the AWJM parameters for machining aluminium/tungsten carbide composites. The main objective of this study is to optimize the AWJM parameters by integrating response surface methodology with ABC algorithm to have an effective solution.

15.2 MATERIALS AND METHODS

Aluminium was reinforced with titanium carbide (2%, 4%, 6%, 8% and 10%) by the stir casting technique. Pure aluminium alloys having 99% purity in the form of rod was melted at 800°C using an electric resistance furnace. Preheated TiC was added into the melt at a stirring speed of 500 rpm. One weight percent of Mg was added with the melt to improve the wettability of reinforcement particles. The melt was then poured into the mould cavity and allowed to cool by keeping the mould at room temperature, and the composites were cut to 100 mm × 100 mm × 10 mm sizes (Figure 15.1).

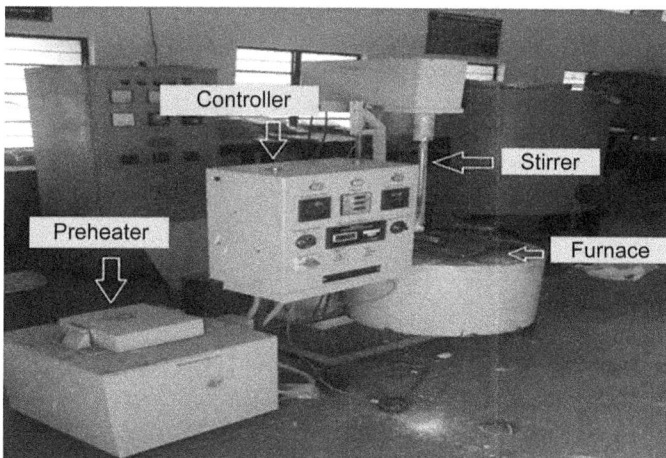

FIGURE 15.1   Stir casting set-up.

An OMAX model abrasive water jet machine was used for conducting experiments. The machine is equipped with a gravity feed-type abrasive hopper, an abrasive feeder system, pneumatically controlled valves, and a work table with dimensions of 3000 mm × 1500 mm. An injection-type (sapphire) nozzle of diameter 0.28 mm in the machine was used to deliver the water at a peak pressure of 3600 bar, and the equipment was controlled by a PLC-based system. The levels of AWJC parameters, namely the traverse speed, stand-off distance and percentage TiC, were selected based on the existing literature. The abrasive garnet particle of 80 mesh size was used in this study. Surface roughness was measured using a Taylor Hobson surface roughness tester. The material weight was measured by a precision balance having the least measurement of 0.001 g. The material removal rate (MRR) is calculated using Eq. (15.1).

$$MRR = \left(\text{Weight before machining} - \text{Weight after machining}\right) / \text{Time taken} \quad (15.1)$$

15.3   RESULTS AND DISCUSSION

Response surface methodology is an empirical method used to obtain the relation between the process parameters of AWJM and their corresponding response variables. A second-order polynomial linear regression equation to represent the AWJM parameters can be developed using Eq. (15.2).

$$y = a_o + \sum_{i=1}^{n} a_i x_i + \sum_{i=1}^{n} a_{ii} x_i^2 + \sum_{i<j}^{n} a_{ij} x_i x_j + \varepsilon \quad (15.2)$$

where y is the response to the input variables, x_i and x_j designate the input parameters, and $x_i x_j$ and x_i^2 symbolize the interaction and quadratic structure of the inputs. a_i, a_{ii} and a_{ij} are the coefficients of regression, and ε is the error. Response surface model to estimate the MRR and SR was developed with the help of Design–Expert software. The process parameters considered for this study are listed in Table 15.1. A full factorial response surface model with 20 sets of experiments as shown in Table 15.2 was used in this study. Experiments were conducted as per the table, and mathematical models were developed to estimate the relation between the output parameters and its input variables. Mathematical models developed to calculate the material removal rate and surface roughness are shown in Eqs. (15.3) and (15.4).

TABLE 15.1
Input Process Parameters and Their Values

Parameter	Unit	Symbol	Value				
Stand-off distance	mm	SOD	2	4	6	8	10
Traverse speed	mm/min	TS	100	125	150	175	200
%TiC	wt%	TiC	0	2	4	6	8

TABLE 15.2
Input Process Parameters and Outputs

S. No.	Stand-off Distance (mm)	Traverse Speed (mm/min)	% TiC (wt%)	Material Removal Rate (mm³/min)	Surface Roughness (μm)
1	10	150	4	1.69	2.54
2	4	125	0	1.939	2.236
3	6	150	0	2.402	2.145
4	8	175	0	2.557	3.193
5	10	200	0	2.991	3.456
6	2	100	2	1.672	1.915
7	4	125	2	1.91	2.418
8	6	150	2	2.111	2.797
9	8	175	2	2.557	2.799
10	10	200	2	2.945	2.99
11	2	100	4	1.661	2.427
12	4	125	4	2.107	2.873
13	6	150	4	2.25	2.307
14	8	175	4	2.522	2.628
15	10	200	4	2.922	3.09
16	2	150	4	2.816	2.67
17	4	125	6	1.872	1.866
18	6	150	6	2.373	2.504
19	8	175	6	2.359	2.477
20	6	150	8	1.98	3.101

$$MRR = -0.359 - (0.109 * SOD) + (0.0219 * TS) + (0.0600 * TiC) - (0.0034 * SOD^2)$$

$$+ (0.000019 * SOD * TS) - (0.00607 * TiC^2) - (0.00637 * SOD * TiC)$$

$$(15.3)$$

$$SR = 1.17 - (0.007 * SOD) + (0.00790 * TS) + (0.081 * TiC) + (0.0065 * SOD^2)$$

$$+ (0.000005 * SOD * TS) + (0.0103 * TiC^2) - (0.0239 * SOD * TiC) \qquad (15.4)$$

Analysis of variance (ANOVA) is performed to estimate the significance of the developed model. ANOVA table for estimating the significance of the developed model to estimate the MRR and SR is shown in Table 15.3. The p-value of both MRR and SR is less than 0.0001 and is said to be significant. Normally, it is expected that for a model to be significant, the p-value has to be less than 0.05. The F-value as per the standard F-chart for a degree of freedom of 7 for the model and 12 for the residual is 3.012. For the model to be significant, it is expected that the tabulated F-value should be higher than the standard F-value. It can be observed that the F-values for MRR and SR are 42.92 and 12.52, acknowledging the model to be significant.

TABLE 15.3
Analysis of Variance for MRR and SR

Source	Sum of Squares	df	Mean-Square	F-Value	p-value Prob > F
		Material removal rate			
Model significant	3.42	7	0.49	42.92	< 0.0001
Residual cor total	0.14	12	0.011	3.56	19
Std. Dev. mean	0.11	R-squared 0.9616	2.28	Adj R-squared 0.9392	
C.V. %	4.68 Pred R-squared	N/A			
Pressure	N/A Adeq precision	19.2410			
		Surface roughness			
Model significant	2.02	7	0.29	12.52	< 0.0001
Residual cor total	0.38	12	0.11	2.40	19
Std. Dev. mean	0.14	R-squared 0.81948	2.62	Adj R-squared 0.7785	
C.V. %	6.93 Pred R-squared	N/A			
Pressure	N/A Adeq precision	6.261			

Adequate precisions for the models are 19.241 and 6.261 and are greater than 4, indicating the model to be significant. The R-square values of 0.9616 and 0.81948 represent the accuracy of the model at 95% level of confidence. The R-square values of MRR and SR are 0.9128 and 0.8066, respectively. The adequate precision values of MRR (10.965) and SR (8.435) are greater than 4, stating the adequacy of the noise signal. Figure 15.2 representing the relation between the predicted and actual values shows that data points are close to the centre line. This indicates the closeness of the predicted value to the actual value. Based on all factors, it can be considered that the developed models are accurate at 95% confidence level.

15.3.2 EFFECT OF INPUT PARAMETERS ON MRR

It can be observed from Table 15.2 that the material removal rate varies from 1.872 (min. value) to 2.991 mm³/min (max. value). According to Eq. 2, the stand-off distance (0.109) is the prime factor influencing the material removal rate, followed by traverse speed (0.0219) and percentage TiC (0.060). The material removal rate is also influenced by significant parameters, namely square of stand-off distance, %TiC, interaction between stand-off distance and traverse speed, and interaction between stand-off distance and %TiC. It can be evidenced that as the stand-off distance increases from 2 to 10 mm, the material removal rate of the composites decreases (Figure 15.3a). On the contrary, an increase in traverse speed from 100 to 200 (mm/min) increases the material removal rate. An increase in TiC from 0% to 8% by weight decreases the material removal rate of the composites (Figure 15.3b). Similar results were witnessed by other researchers also [6, 8]. An increase in %TiC increases the resistance to erosion by the abrasive particles in the high-pressure abrasive jet.

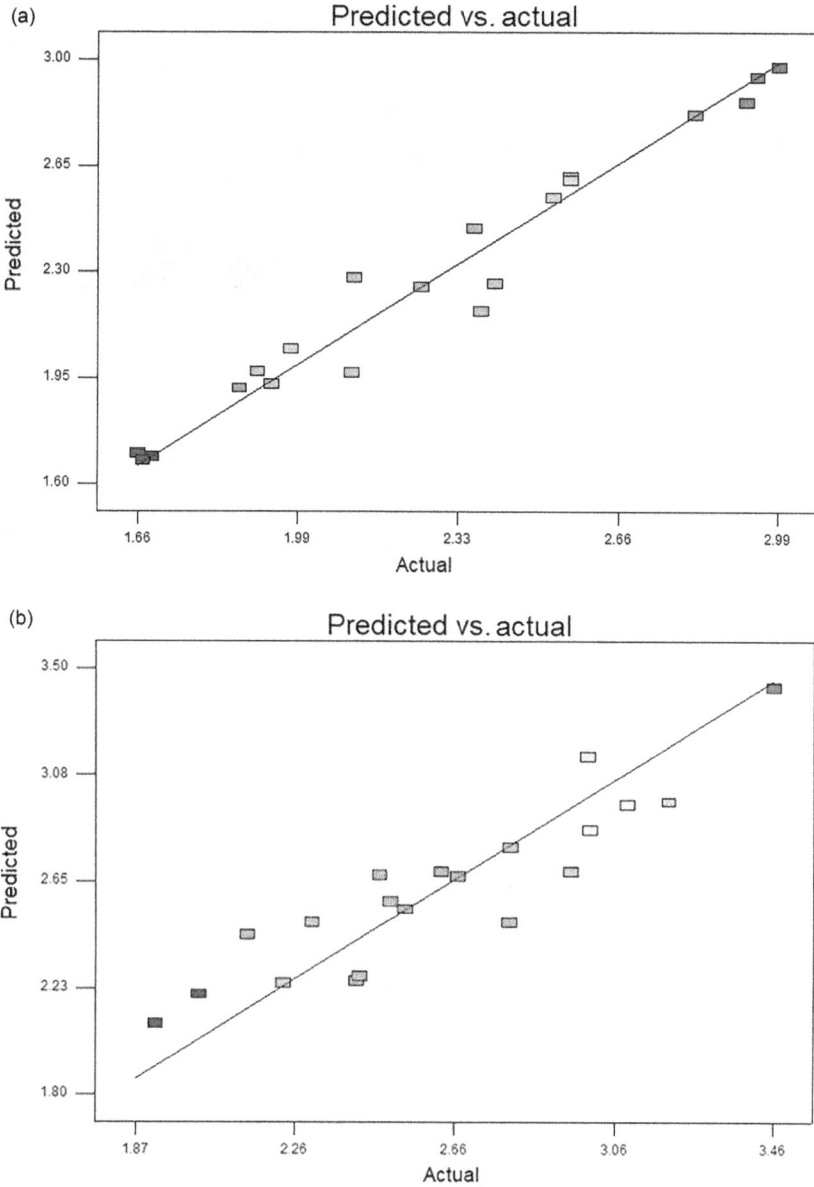

FIGURE 15.2   Predicted value vs actual value: (a) MRR; (b) SR.

This is possible due to the increase in hardness and also due to the strong bonding between the aluminium matrix and the TiC particles. An increase in the stand-off distance normally decreases the efficiency of the jet striking the substrate due to dispersion and leads to reduced concentration of abrasive jet. An increase in traverse speed increases the kinetic energy of the high-pressure abrasive along the surface and covers a large portion of the substrate and increases the material removal rate.

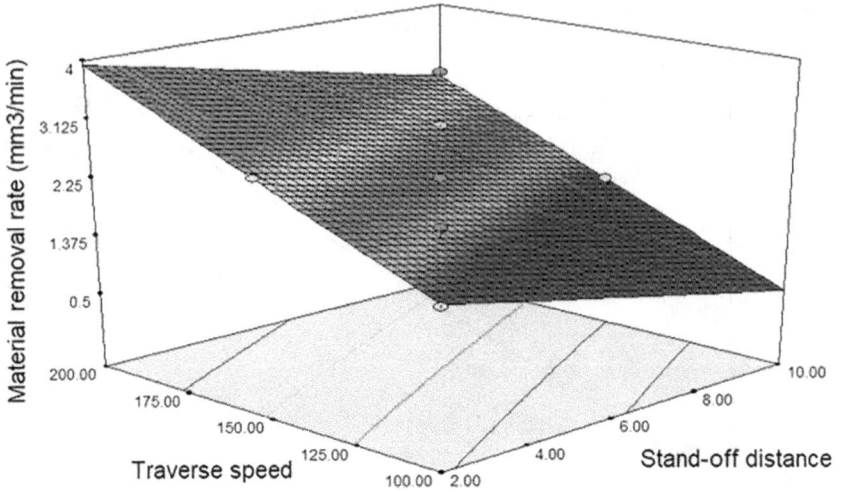

FIGURE 15.3A Influence of input parameters on MRR: traverse speed and stand-off distance vs MRR.

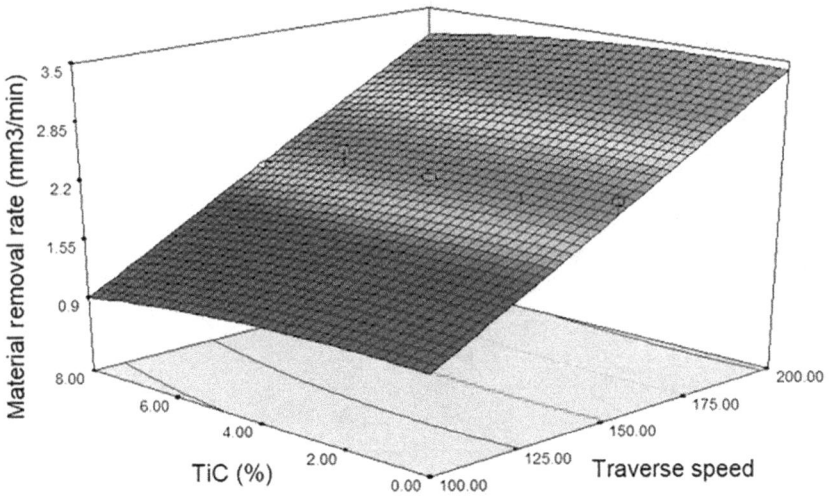

FIGURE 15.3B Influence of input parameters on MRR: traverse speed and %TiC vs MRR.

15.3.3 EFFECT OF INPUT PARAMETERS ON SR

It can be observed from Table 15.2 that the surface roughness of the composites ranges from 1.866 (min. value) to 3.456 µm (max. value). According to Eq. (15.3), percentage TiC (0.081) is the principal factor influencing the surface roughness, followed by traverse speed (0.00790) and stand-off distance (0.0070). Surface roughness is also influenced by significant parameters, namely square of stand-off distance, %TiC, interaction between stand-off distance and traverse speed, and interaction between stand-off distance and %TiC. An increase in traverse speed from 100 to 200 (mm/min) increases the surface roughness of the composites (Figure 15.4a). As the stand-off distance increases from 2 to 10mm, the SR of the composite decreases and, towards the end, there was a mild increase in the surface roughness. An increase in the presence of TiC in the aluminium matrix increases the SR of the composites (Figure 15.4b). Similar results were witnessed by other researchers also [6–8]. An increase in traverse speed covers a major surface of the composite

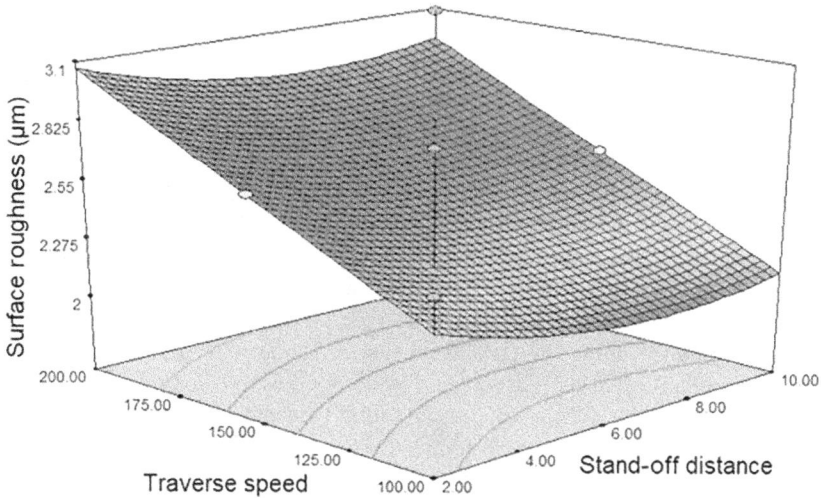

FIGURE 15.4A Influence of input parameters on SR: traverse speed and stand-off distance vs SR.

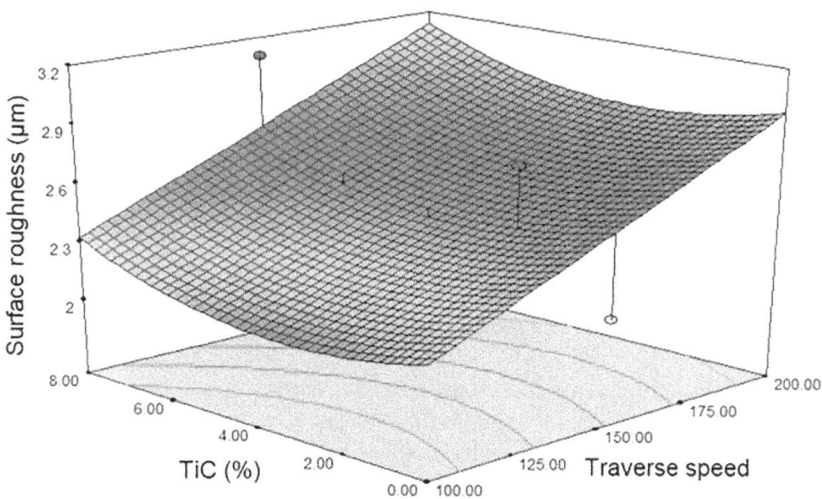

FIGURE 15.4B Influence of input parameters on SR: traverse speed and %TiC vs SR.

at a faster rate. This leads to irregular removal of the composite materials, leaving some uncut particles in the form of debris along the surface and increasing the surface roughness. At a smaller stand-off distance, the high-energy abrasive particles and the eroded chip particles stay back down the surface of the composites due to the high concentration of the jet and increase the surface roughness. An increase in the stand-off distance decreases the jet energy and decreases the surface roughness. An increase in TiC particles increases the resistance to erosion and remains along the surface of the composites in the form of pull-outs and increases the surface roughness.

15.4 BEE COLONY ALGORITHM

Karaboga and Basturk [20] developed a nature-inspired algorithm called artificial bee colony (ABC) algorithm that depends on the intelligent behaviour of a honey bee swarm. According to this algorithm, the colony comprises of three groups of artificial bees, namely employed bees, onlookers and scouts. The employed bee identifies the food source (nectar information) and reaches back to the hive and dances based on the area. The onlookers follow the dance of the employed bee and identify the food source. Once the food source becomes deserted, the employed bee becomes scout and begins searching for a new food source. In the ABC algorithm, the probable solution for optimization is designated by the food source and the quality of ABC algorithm relies on the nectar amount. The intelligence of ABC algorithm includes three essential components, namely food sources, employed foragers (employed bee) and unemployed foragers (scouts and onlookers). It is always considered that the number of food sources will be equal to the number of employed bees. The two control parameters involved in ABC algorithm are the colony size N and the number of global cycle numbers M. The main steps of the algorithm are given below:

    Initial food sources are produced for all employed bees
    REPEAT
- The employed bee reaches the food source and decides upon the nectar quantity and dances in the hive.
- The onlooker bee follows the dance of the employed bee, and depending on the dance of the employed bee, it moves to the source and determines the nectar amount after choosing a neighbour around.
- The food sources are deserted.
- The scouts are sent to discover new food sources.
- The best food source is memorized.

    UNTIL (requirements are met)

In the ABC algorithm, every food source stand is a possible solution to the problem that is under consideration. The set initialization $[x_i = (x_{i1}, x_{i2}, \ldots, x_{iD})]$ is generated by the scout bees, and the ith food source is given by Eq. (15.4). x_i is an S-dimensional vector, where S indicates the number of design variables [25].

$$x_{ij} = lb_j + \text{rand}(0,1)\left(ub_j - lb_j\right) \qquad (15.5)$$

where $i = 1, ..., N/2, j = 1, ..., D$ is the vector representation of the food source location, and the upper and lower bounds are represented by ub_j and lb_j, respectively, of the jth parameter.

After initialization, the employed bee repeats the search process cycle and one employed bee is connected with only one food source; the new search is given in Eq. (15.6):

$$v_{ij} = x_{ij} + \phi_{ij}\left(x_{ij} - x_{kj}\right) \tag{15.6}$$

where ϕ_{ij} represents a random number in $[-1, 1]$, x_i indicates the current food source, v_i represents the new food source, and x_k is a neighbour food source. The k value can be randomly chosen from $k = 1$ to $N/2$ and is different from i.

Once the employed bees complete the search and the onlooker bees choose food source based on the nectar amount, then the probability of i can be expressed as Eq. (15.7):

$$P_i = \frac{\text{fit}_i}{\sum_{i=1}^{N/2} \text{fit}_i} \tag{15.7}$$

where fit_i represents the current food source i and is equivalent to the nectar amount. Once the food source is deserted, the scout bee generates a new food source. This cycle is repeated until cycle numbers $= M$.

15.4.1 Proposed Modified ABC (MABC) Algorithm

The basic ABC algorithm was originally designed for continuous function optimization. To optimize the abrasive water jet machining parameters of Al/TiC, some modifications are required in the basic ABC algorithm. The primary objective is to identify the set of input parameters that produces maximum MRR and minimum SR. The computational procedure of the proposed modified ABC algorithm (MABC) is explained below.

15.4.2 Computational Procedure of the Proposed MABC Algorithm

Step 1: The response surface methodology is employed to design the experiments.
Step 2: Use the developed mathematical model for simulation.
Step 3: Initialize the population: $S = \{s_1, s_2, ..., s_{PS}\}$ with the boundary.
  Stand-off distance (mm) $- 2 \leq \text{SOD} \leq 10$
  Traverse speed (mm/min) $- 100 \leq \text{TS} \leq 200$
  %TiC (wt%) $- 0 \leq \text{TIC} \leq 8$
  $N = 40$ (no. of food sources)
  $M = 500$ (no. of cycles)
  Limit $= 10$ (no. of trials)
Step 4: Evaluate each solution using the developed RSM model.
Step 5: Employed bee phase:
  Produce new food source positions, $S_i^* = \text{SOD, TS, TiC}$ in the neighbourhood of S_i.

For $i = 1, 2, \ldots, N$, repeat the subsequent sub-steps:

Develop a new solution S_i^* for the ith employed bee that is associated with solution S_i by using the following four neighbouring approaches;

1. Perform one insert operator to a solution S_i.
2. Perform two insert operators to a solution S_i.
3. Perform one swap operator to a solution S_i.
4. Perform two swap operators to a solution S_i.

One of the above-said neighbouring approaches is selected randomly to produce and evaluate the new solution S_i^*. If S_i^* is better than or equal to S_i, let $S_i = S_i^*$.

Step 6: Onlooker phase:

For each $i = 1, 2, \ldots, N$, repeat the subsequent sub-steps:

- Select a food source from the population for the onlooker bee S_i by using the tournament selection.
- Generate a new solution S_i^* for the onlooker by using the same method as used by the employed bee to develop a new solution and evaluate it.
- If S_i^* is better than or equal to S_i, let $S_i = S_i^*$.

Step 7: Scout phase:

For each $i = 1, 2, \ldots, N$, repeat the following sub-steps:

- If a solution S_i in the population does not improve during the final trial, then it is abandoned.
- Create a new onlooker solution using the same strategy used by the employed bee to produce a new solution on the best solution.

Step 8: Store the best solution achieved so far.

Step 9: If the termination criterion is reached, return the best solution found so far; otherwise, go to Step 5.

Based on the above approach, the combined response surface/bee colony optimization was carried out. The main objective of this optimization is to maximize the material removal rate and to minimize the roughness along the surface of the composites. The optimized values of the modified approach are given as follows:

Stand-off distance: 9 (mm)
Traverse distance: 165 (mm/min)
Percentage TiC: 0.9 (wt%)
Material removal rate: 2.024 (mm³/min)
Surface roughness: 2.832 (µm)

15.5   CONCLUSIONS

In this study, AWJM parameters of aluminium/TiC composites were optimized by a modified RSM-ABC algorithm and the following conclusions are drawn:

- Aluminium/TiC composites were fabricated by the stir casting technique.
- RSM was utilized to the study the MRR and SR of AWJM by varying the stand-off distance, traverse speed and TiC.

- The significance of the developed models was checked by ANOVA, and the developed models were accurate within the specified ranges. The R-square values observed for MRR and SR are 0.9616 and 0.81948, respectively, indicating the accuracy of the model.
- The stand-off distance is the prime factor influencing the material removal rate, followed by traverse speed and percentage TiC. An increase in MRR is due to the increase in the kinetic energy of the high-pressure abrasive particles hitting the workpiece.
- The %TiC is the principal factor influencing the surface roughness, followed by traverse speed and stand-off distance. Scattering of the surface jet along the surface plays a vital role in influencing the surface roughness.
- AWJM parameters were optimized by integrating response surface methodology with ABC algorithm to have an effective solution.
- The optimum material removal rate and surface roughness in this study are 2.024 (mm^3/min) and 2.832 (μm), respectively, with the following process parameters: stand-off distance 9 (mm), traverse speed 165 (mm/min) and TiC 0.9 (wt%).

REFERENCES

1. Bagchi, A., Srivastava, M., Tripathi, R., Chattopadhyaya, S., 2019. Effect of different parameters on surface roughness and material removal rate in abrasive water jet cutting of Nimonic C263, *Mater. Today: Proc.* doi:10.1016/j.matpr.2019.09.104.
2. Patel, D., Tandon, P., 2015. Experimental investigations of thermally enhanced abrasive water jet machining of hard-to-machine metals, *CIRP J. Manuf. Sci. Technol.* 10, 92–101.
3. Nag, A., Scucka, J., Hlavacek, P., Klichova, D., Srivastava, A.Kr., Hloch, S., Dixit, A.R., Foldyna, J., Zelenak, M., 2018. Hybrid aluminium matrix composite AWJ turning using olivine and Barton garnet, *Int. J. Adv. Manuf. Technol.* 94, 2293–2300.
4. Santhanakumar, M., Adalarasan, R., Rajmohan, M., 2016. Parameter design for cut surface characteristics in abrasive waterjet cutting of Al/SiC/Al$_2$O$_3$ composite using grey theory based RSM, *J. Mech. Sci. Technol.* 30 (1), 371–379.
5. Madhu, S., Balasubramanian, M., 2018. Impact of nozzle design on surface roughness of abrasive jet machined glass fibre reinforced polymer composites, *Silicon* 10, 453–2462.
6. Gnanavelbabu, A., Rajkumar, K., Saravanan, P., 2018. Investigation on the cutting quality characteristics of abrasive water jet machining of AA6061-B4ChBN hybrid metal matrix composites, *Mater. Manuf. Process.* 33 (12), 1313–1323.
7. Rethan Raj, R., Kanagasabapathy, H., 2018. Influence of abrasive water jet machining parameter on performance characteristics of AA7075-ZrSiO$_4$-hBN hybrid metal matrix composites, *Mater. Res. Express*, 5 (10). doi:10.1088/2053-1591/aadabf.
8. Ravi Kumar, K., Sreebalaji, V.S., Pridhar, T., 2018. Characterization and optimization of Abrasive Water Jet Machining parameters of aluminium/tungsten carbide composites, *Measurement* 117, 57–66.
9. Veerappan, G., Ravichandran, M., 2019. Experimental investigations on abrasive water jet machining of nickel-based superalloy, *J. Braz. Soc. Mech. Sci. & Eng.* 41, 528.
10. Ahmed, T.M., El Mesalamy, A.S., Youssef, A., El Midany, T.T., 2018. Improving surface roughness of abrasive waterjet cutting process by using statistical modeling, *CIRP J. Manuf. Sci. Technol.* 22, 30–36.

11. Natarajan, Y., Murugesan, P.K., Mohan, M., Khan, S.A.L.A., 2020. Abrasive Water Jet Machining process: A state of art of review, *J. Manufac. Process* 49, 271–322.

12. Srivastava, A.K., Nag, A., Dixit, A.R., Tiwari, S., Srivastava, V.S., 2019. Hardness measurement of surfaces on hybrid metal matrix composite created by turning using an abrasive water jet and WED, *Measurement* 131, 628–639.

13. Ravi Kumar, K., Mohanasundaram, K.M., Arumaikkannu, G., Subramanian, R., 2012. Artificial neural networks based prediction of wear and frictional behaviour of aluminium (A380)–fly ash composites, *Tribol. Mater. Surf. Interfaces* 6(1), 15–19.

14. Manoj, M., Jinu, G.R., Muthuramalingam, T., 2018. Multi response optimization of AWJM process parameters on machining TiB_2 particles reinforced Al7075 composite using taguchi-DEAR methodology, *Silicon* 10, 2287–2293.

15. Balamurugan, K., Uthayakumar, M., Sankar, S., Hareesh, U.S., Warrier, K.G.K., 2019. Predicting correlations in abrasive waterjet cutting parameters of Lanthanum phosphate/Yttria composite by response surface methodology, *Measurement* 131, 309–318.

16. Ravi Kumar, K., Sreebalaji, V.S., 2015. Desirability based multiobjective optimization of abrasive wear and frictional behavior of aluminium (Al/3.25Cu/8.5Si)/fly ash composites, *Tribol. Mater. Surf. Interfaces* 9(3), 128–136.

17. Ravi Kumar, K., Soms, N., 2019. Desirability-based multi-objective optimization and analysis of WEDM characteristics of aluminium (6082)/tungsten carbide composites, *Arab. J. Sci. Eng.* 44, 893–909.

18. Jagadish, Bhowmik, S., Ray, A., 2016. Prediction and optimization of process parameters of green composites in AWJM process using response surface methodology, *Int. J. Adv. Manuf. Technol.* 87, 1359–1370.

19. Yadav, S.L., Phogat, M., 2017. Study of nature inspired algorithms, *Int. J. Comput. Trends Technol. (IJCTT)* 49 (2), 100–105.

20. Karaboga, D., Basturk, B., 2007. A powerful and efficient algorithm for numerical function optimization: artificial bee colony (ABC) algorithm, *J. Glob. Optim.* 39, 459–471.

21. Karaboga, D., Gorkemli, B., Ozturk, C., Karaboga, N., 2014. A comprehensive survey: Artificial bee colony (ABC) algorithm and applications, *Artif. Intell. Rev.* 42, 21–57.

22. Farooq, M.U., Salman, Q., Arshad, M., Khan, I., Akhtar, R., Kim, S., 2019. An artificial bee colony algorithm based on a multi-objective framework for supplier integration, *Appl. Sci.* 9, 588. doi:10.3390/app9030588.

23. Rao, R.V., Pawar, P.J., Shankar, R., 2008. Multi-objective optimization of electrochemical machining process parameters using a particle swarm optimization algorithm, *Proc. Inst. Mech. Eng. B* 222, 949–958.

24. Solaiyappan, A., Mani, K., Gopalan, V., 2014. Multi-objective optimization of process parameters for electrochemical machining of 6061Al/ 10%Wt Al_2O_3/ 5%Wt SiC composite using hybrid fuzzy-artificial bee colony algorithm, *Jordan J. Mech. Indus. Eng.* 8 (5), 323–331.

25. Zhou, J., Zhang, X., Zhang, G., Chen, D., 2015. Optimization and parameters estimation in ultrasonic echo problems using modified artificial bee colony algorithm, *J. Bionic Eng.* 12, 160–169.

26. Yildiz, A.R., 2013. A new hybrid artificial bee colony algorithm for robust optimal design and manufacturing, *Appl. Soft Comput.* 13, 2906–2912.

Index

For Product Safety Concerns and Information please contact our EU
representative GPSR@taylorandfrancis.com
Taylor & Francis Verlag GmbH, Kaufingerstraße 24, 80331 München, Germany

www.ingramcontent.com/pod-product-compliance
Lightning Source LLC
Chambersburg PA
CBHW060348220326
41598CB00023B/2847

9 7 8 0 3 6 7 5 3 2 6 1 1